ONCE UPON ATARI

HOW I MADE HISTORY BY KILLING AN INDUSTRY

HOWARD SCOTT WARSHAW

Copyright ©2020 Howard Scott Warshaw, MA, ME, LMFT: (California MFT #52529)

Without limiting the rights under copyright reserved above, no part of this publication may be reproduced, stored in or introduced into retrieval system, or transmitted, in any form or by any means (electronic, mechanical, photocopying, recording, or otherwise), without the prior written permission of both the copyright owners and the publisher of this book.

Published By: SCOTT WEST Productions

Book Design By: Tommy Owen (TommyOwen.com)

ISBN #: 978-0-9862186-6-8

Version 1.0

Printed in the United States of America All rights reserved.

All images & illustrations were either licensed, created, or permission was granted to use them.

TABLE OF CONTENTS

INTRO

Dedication	-3
Foreword	-2

CHAPTERS

Chapter 1	Lightning Strikes	1
Chapter 2	King Learjet	6
Chapter 3	Endings & Beginnings	15
Chapter 4	Here Comes The Crash	52
Chapter 5	Are You Game?	72
Chapter 6	What's New?	100
Chapter 7	A Nerd World Country	108
Chapter 8	Games & Their Makers	122
Chapter 9	Conflicting Feelings	136
Chapter 10	Do it in 5 weeks	152
Chapter 11	So Many Questions	165
Chapter 12	Lump In The Snake	182
Chapter 13	The Home Stretch	196
Chapter 14	Kudos, Burnout & Letting Go	213
Chapter 15	Path Connected	230
Chapter 16	Fiddling While ROMs Burn	246
Chapter 17	Who'll Stop The Rain	253
Chapter 18	Tough Act To Follow	273
Chapter 19	Sweeter The 2nd Time	283
Chapter 20	Ground Breaking	294
Chapter 21	Impact	304
Chapter 22	Afterglow	314

HERE'S WHAT THEY'RE SAYING ABOUT ONCE UPON ATARI:

"Once Upon Atari is, ostensibly, a book about nostalgic videogames; a tale set in a golden era that for many of us represents the dawn of an art form that has irrevocably and powerfully changed our lives and our world. And, as a personal history of these times, it is a moving and emotional trip, filled with interesting details and stories as told by someone so deeply inside them as to have been a part of their essential fabric. But that's not the soul of this book, which is, in its true heart, the journey of a man – a programmer of immense talent – sucked in by a machine of greed and spat out, carelessly, after giving all and everything to it. A journey of learning and redemption, set against a world all about the seemingly magical, effortless creation of fun and joy. This wonderful book is the capstone of that journey, and to read it is to be a part of that odyssey, and to partake in the lessons that it challenges us to learn."

- **Seamus Blackley, Father of the XBox**

"Howard sheds light on Atari's most tumultuous period. If you want to understand the true story of the video game crash, I highly recommend this book."

- **Nolan Bushnell, Atari Founder**

"In those few years that Howard was at Atari, he experienced more than most of us ever will over decades of working in this crazy field. All of the extreme highs and deep lows were condensed into a short exhaustive burn that's both beautiful and serves as an allegory for the pitfalls of unchecked creative sacrifice. This book is just as much for those interested in the early days of video games as it is for anyone giving their all to any kind of creative endeavor. He looks back from the vantage point of an accomplished therapist, not with any warning or overt guidance, but with the contagious excitement of someone who is content with the role he's played, and optimistic about the future of an industry he helped create. I've worked in games for over twenty-five years, and I've known Howard for most of them. I can honestly say that he represents the heart and soul of our industry more than anyone else I know."

- Mike Mika, Video Game Developer, Author and Historian

"*Once Upon Atari* brings me right back to my childhood video game days, living in Silicon Valley where it was all unfolding. Howard pulls back the curtain on an industry that exploded into our family rooms with his help, and shares incredible behind the scenes stories of how that curtain came crashing down in the most unlikely place. A must-read."

- Bret Burkhart, Broadcaster, KGO Radio

October of 2016, outside our favorite restaurant.

Atari office in 1982

DEDICATION

This book is dedicated to my good friend, Jerome Domurat.
Jerome was a technologist in the world of art just as I was an artist in the world of technology. We were each comfortably out of place in our respective environments. Two mirrors facing each other, reflecting endlessly upon the universe we shared. You may not get to read it, but I'm glad we lived it together. And good luck with that dPhone!

For my wife, Sherri. She's responsible for some of the best lines you'll see and some of the worst ones you won't. She matches my creativity yet brings calm to my chaos, for which I am eternally grateful. My relationship with Sherri started about the same time as my journey to become a psychotherapist. Life attempted to block each early on, but neither was daunted. Both continue to be amazing.

To Bob Saenz. I'll tell you about him later.

Thank you to Emily Boykin of Em and a Pen, whose keen editorial eye improved the trip.

Thank you to Tommy Owen (www.TommyOwen.com), whose creative design skills improved the destination.

Thank you to Dave Staugas, not only a great programmer, but a great photographer and prognosticator. He shot the cover for this book decades before I knew there would be a book.

And a special thank you to Tod Frye, it simply would not have been Atari without you.

FOREWORD

There's an old adage that you should avoid meeting one of your heroes, because the chances are that you'll end up being disappointed somehow. Thankfully, I've never found this to be the case. I've been lucky enough to meet several of my heroes in person, and to even work with a few of them. Never once have I been disappointed. Quite the opposite. I've always ended up admiring the person even more than I did previously, when I just knew them through their creative output. A few of my childhood heroes have even become personal friends. Howard Scott Warshaw is one of them.

I grew up in the 1970s and 80s in a small town in Ohio, and two of my childhood passions were movies and video games. That was how Howard first entered my life. He created one of my favorite home video games, *Yars' Revenge* for the Atari 2600, which was the first game ever to come with a comic book that laid out its back story. Howard also created the first video game adaptation of a movie, *Raiders of the Lost Ark*. Raiders was one of my all-time favorite films, and Howard's adaptation became one of my all-time favorite games. Can you imagine? A video game that allowed a regular kid to become Indiana Jones while he was sitting cross-legged in front of his TV on Saturday morning! This was something completely new, and it changed my whole concept of what a video game could be.

Howard is probably most well-known for also being the creator of *E.T.*, another game based on a Steven Spielberg movie. I got this game for Christmas in 1982 and I absolutely adored it! It was the most detailed and advanced Atari 2600 game I had ever played. It contained an entire cube-shaped world for the player to explore, treasures to find, enemies to avoid, and even an ally, in the form of E.T.'s human pal, Elliott, who would show up to help you out when things weren't going your way. I happily played it for hours and hours. It was the first game that contained multiple Easter eggs for the player to find. It was also the first video game that ever made me tear up, when I managed to help E.T. successfully phone home for the first time…

Years later, when the Internet came along and I saw people attempting to label *E.T.* as the "worst video game of all time," I knew they were all completely full of crap. And I still do. *E.T.* was one of the most innovative and groundbreaking video games of its time. Like Howard's two previous games, it was filled with things that no programmer had ever thought to put into a game before. And best of all, it wasn't an easy game to beat. It was incredibly challenging, and it required you to use your brain, during an era when most weren't and didn't. I'm so grateful that Howard had the courage to try something different, along with the programming genius to pull it off.

I'm also grateful that Howard decided to write about his time working at Atari, so that we can all share his unique experience vicariously. If you grew up playing Atari games like I did, you probably spent a great deal of time wondering what it would be like to visit Atari's headquarters in Sunnyvale, CA, and to meet all of the incredible people who created your favorite games. Thanks to the book you now hold in your hands, you no longer have to wonder.

There's an episode of the Simpsons where Bart visits the offices of MAD Magazine, hoping to get a tour. The receptionist turns him away, saying "Listen, kid. You probably think lots of crazy stuff goes on in there. But this is just a place of business." Bart is about to leave, dejected, but then a door opens and he catches a thrilling glimpse of Alfred E. Neuman and the rest of the MAD writing staff, engaged in all sort of craziness. The place is a nuthouse—just like he'd always hoped and imagined it would be. That's how I felt while reading this book. It's like finding a golden ticket wrapped around a candy bar and winning a guided tour of the chocolate factory given by Willie Wonka himself.

I hope you all enjoy reading about Howard's wild adventure as much as I did.

Ernest Cline | Austin, Texas | November 19, 2020

CHAPTER 0

"You can ignore reality, but you cannot ignore the consequences of ignoring reality."

Ayn Rand

CHAPTER 1
LIGHTNING STRIKES

THE STORM BEFORE THE CALM

"Ouch!"

Airborne grains of sand and flying bits of old trash are pelting me without mercy. Honestly, I never imagined I'd find myself here...

I'm standing in the middle of a garbage dump in the New Mexico desert. It's hot. It's LOUD. A huge sandstorm rages all around us. I'm surrounded by hundreds of people from all over the country. We huddle like penguins for protection against the onslaught. There are news people, construction people, food people, film people and even some local politicians, but the vast majority are fans. Classic Video Game fans. People who smile at the mere mention of the word "Atari."

We're all here, braving the heat and the storm, watching huge noisy yellow machines reaching deep into the ground, literally digging up my past right before my eyes. A big yellow arm disappears into a hole, bringing up another claw-bucket of ancient garbage and detritus. The arm swings around and dumps its load before returning for the next scoop, leaving behind a dusty pile of old refuse. The ground between the machines and a thin plastic retaining fence is dotted with such piles. Each one holds the promise of a "nugget." Bodies press against the fence, straining to get a closer look at the latest droppings. "Is it there?" "Can you see one?" Or is this just more ammunition for the relentless gusting winds?

What are we doing here? We're searching for evidence. Specifically, we're hoping to unearth the murder weapon with which I allegedly killed a multibillion-dollar industry back in the early '80s. And as good suspects do, I'm denying its existence. For decades I've said the very idea is ridiculous, but today I really hope I'm wrong. I've explained many times over why this whole operation makes absolutely no sense. But I'd forgotten the cardinal rule:

When you expect things to make sense, you're losing touch with Atari.

This is another remarkable day in my life. I've had many, but this one is special. Saturday, April 26th, 2014 is the longest day of my life, because it started on July 27th, 1982.

THE PHONE CALL

On the afternoon of Tuesday, July 27th, 1982 I'm sitting in my office at 275 Gibraltar Drive, on Atari's main campus in Sunnyvale, California. I'm hanging out with Jerome Domurat after putting the final touches on Raiders of the Lost Ark, the longest development of all my games. Jerome is my graphics/animation designer and my good friend. We're having fun in our usual way, taking turns reading aloud from National Lampoon magazine's letters to the editor, when a call comes in: "Will you please hold for Ray Kassar?"

Will I hold for Ray Kassar? The Chief Executive Officer of Atari? My boss's boss's boss's boss's boss? The guy who signs my paychecks? "Yes, I'll hold for him."

A phone call from Ray Kassar is a very unusual thing in my experience. However, this is not my first time chatting enjoyably with our CEO. The first time was at a press event. I was demo-playing my first game, Yars' Revenge, on one of the first ever Big-Screen TVs (a hulking rear-projection monstrosity). Ray emerged from the slew of media people crawling around the room. He approached me and said, "Hello Howard, I heard about what you did with Yars."

"Yeah? What did you think about that, Ray?"

He half-smiled, "Just keep making games, Howard." Then he turned and melted back into the traffic. That was my first encounter with Ray Kassar. The last time we met, however, was a bit more memorable…

Roughly two months before answering this phone call, I was nearing the final stages of development on Raiders of the Lost Ark, the first ever video game based on a movie. It was a dog-and-pony day, which means key execs are cruising engineering for demos (somewhat akin to visiting the zoo) and we show the current state of our games to anyone being escorted by our bosses. I take game demos pretty seriously, but this time was special. The man himself, Ray Kassar, was coming down from on high to take the tour. He had his entourage in tow, including extras from marketing, legal and the odd vice president or two. You knew when Ray was coming because his distinctive cologne always preceded him. He came wafting in and took the guest chair while the others stood around him like a halo of nodding assent. I had the game ready to go and Tchaikovsky's "Overture of 1812" (the one with the cannons) cued up on the office stereo. It lends an impressive ambiance to the demonstration, well beyond the capabilities of my development station.

[NOTE to the Non-Nerd: A Development Station (or Dev Station) is a specialized piece of hi/low-tech computer hardware (frequently tucked into a black metal box) where game programmers can test-run and debug their software in a reliable environment. It is designed to prevent programmers from having anything other than themselves to blame for their product issues. Of course, this design goal is not always realized.]

I press Play on the stereo, pick up my game controller and roll through the demo. Ray offers occasional comments, each of which is quickly and enthusiastically affirmed by the entourage.

Now it isn't every day I get Ray Kassar in my office, so being the braying ass I'm given to be at times in my mid 20's, I took the opportunity to share some thoughts and suggestions (read: criticisms & complaints) as to how the company might be better run. Mouthing off to the big man is not usually the smartest strategy, but it's easier when your work represents a significant chunk of corporate profits, past and future.

After sitting politely through a more-than-reasonable bit of this, Ray cuts in and says, "Interesting ideas. Perhaps we should switch jobs for a day."

Instantly I fire back, "I'm good with that, Ray. Here's my dev station. Just give me your fragrance and let's go."

And the room froze.

Uh-oh, have I gone too far this time? (a question I ask myself all too frequently)

A deafening silence hung there, occasionally broken by stifled chortles. The entourage wants to laugh but they don't want the guillotine. All the king's men were desperately trying to hold their laughter until they got some inkling of Ray's reaction. After what seemed like hours, Ray finally decided to find it amusing and thus unleashed the torrent. Laughter abounded as they shuffled off to the next office.

Since I wasn't fired for that one, I lived to take this call today...

Ray comes on the phone and gets right to the point: "Howard, we need an E.T. game for September 1st. Can you do it?"

Without missing a beat, I say, "Absolutely I can! Provided we reach the right agreement." I know what I mean. Ray knows too. Money.

"That's fine," Ray says, "be at San Jose Airport Thursday morning at 8am. There will be a Learjet waiting to take you to Spielberg's office where you'll present the design for the game."

And there it is. I'm doing the E.T. game! My first thought is: Whoa, I've got 36 hours to do the entire design and prepare a presentation for the fastest video game development ever attempted. My second thought is: Better have a good dinner tonight, it might need to last me a while. And oh yeah, I'm still on the phone...

I assure Ray I'll be fully prepared when I board the Spielberg-express first thing Thursday morning. We say our goodbyes and hang up. This will not be my first encounter with Steven Spielberg. We've met several times before, but this one will require more imagination, creativity and fancy footwork than any other.

I know what I'm actually promising. Games on this system usually take at least 6 months to develop. I'm committing to do one in 5 *weeks*. Am I confident? My hubris is. But right now, I'm already too busy to think about it. Just 36 hours to my first delivery milestone. In order to pull this off, a lot of headwork needs to happen in a very short time. Fortunately, my brain is hard-wired for fast. The tricky part is the balance, staying focused but not tunnel visioned...

Let the thinking begin!

So... where to go for dinner?

CHAPTER 2
KING LEARJET

BACK TO THE DESERT

And now we're back in the New Mexico desert in 2014, because this isn't just a chronicle, it's also a time machine. And a good thing too, because it takes a time machine to understand how that one phone call decades ago began paving a road leading me to this place, this hour, this sandstorm in a dump in the desert.

I woke up this morning several hours earlier and nearly 5,000 feet higher in a mountaintop hideaway hotel, far from this chaos. After a hearty breakfast, we boarded the van of destiny and headed for the Alamogordo city dump. Snaking our way down mountain roads, I was feeling both curious and anxious; Curious about what they'd find under the ground and anxious about what that might mean. Upon arrival, I see something very odd indeed… there's a line of people waiting. A long line. When's the last time you saw hundreds of people standing in line to get into a garbage dump?

I should probably say a bit more about what's going on here. Today, Lightbox & Fuel Entertainment (Hollywood production companies), Xbox Entertainment Studios (a small part of a *huge* corporate entity) and the city of Alamogordo, New Mexico are jointly hosting (and filming) a modern archeological event. This is an excavation (or "dig") to literally uncover the truth behind an enduring urban myth. Specifically, that decades ago Atari trucked millions of unsold E.T. video game cartridges into the desert and buried them here in this dump. I'm here too of course, because I did it. I'm the one.

I made the worst video game of all time!

This is not my opinion. This is the conclusion held by many All-Time lists. Go ahead, Google "worst video games of all time" and see what you get. Countless fans and media people remind me of this "fact" regularly. In 1995, New Media magazine said my E.T. game was so bad it single-handedly caused the video game crash of the early '80s, collapsing an industry with revenues approaching four billion dollars.

It was so bad that Atari needed to bury it deep in the desert just to get rid of the stench! At least that's the legend. Snopes.com says it's true. I've always denied it. I'll tell you why...

When a company is hemorrhaging money to the tune of hundreds of millions of dollars and they find themselves sitting on a mountain of worthless inventory, why would they spend even more money to transport, crush, cement over and bury it? That's a very expensive thing to do. Why not recycle the materials to reduce the cost of making new product that might sell? At the very least, you could simply throw open the doors of the warehouse and let people come in and take it all. Why spend big money getting rid of something you believe is worthless? It doesn't make any sense.

As I said before, when you expect things to make sense, you're losing touch with Atari.

Atari was never about making sense. Atari was about making fun. It was about inventing things that never existed before in ways no one had ever imagined. It was not a sensible place; it was an outrageous place. It was an orgy of creativity and innovation, populated by the most engaging, accomplished and eccentric cast of characters I've ever known. Atari was the perfect place at the perfect time for me... but they didn't see it at first.

After a round of interviews Atari rejected me... but I pushed back. I reasoned, argued, and pleaded with Dennis Koble (the hiring manager running my interview process) until he finally agreed to give me a chance (for a probationary period and a significantly smaller salary)(which I gladly accepted). I kept pushing, because on some deep intuitive level I knew Atari would be my home. It was everything I needed for sustenance and growth in my life. I had to be there.

When my time at Atari ended (which it had to, since nothing so imbalanced can remain standing indefinitely) I knew it would never be equaled.

I did finally exceed it, however. After some thirty years of searching, schooling and internships I finally became what I always wanted to be: a psychotherapist. And now, with my life once again supremely satisfying and rewarding, I find myself in the desert getting sandblasted at the end of a long and winding road which began decades ago with a phone call. I'm waiting to see if my past will rise once more. Is my notorious creation poised to jump out of this ever-deepening hole in the desert floor?

HSW AT THE DIG SITE.

I hope it does. It'll make for a much better movie that way. In fact, the prospect of being wrong has never been more appealing. Besides, I always want my games to be groundbreaking in some fashion. Will my third creation finally break ground in a new and most unexpected way? The irony would be delicious. Speaking of which, I'm getting kind of hungry...

YOUR LEARJET AWAITS

I hate getting up early in the morning. Aside from a brief stint in commercial real estate, I've always worked hard to maintain a life that never needs an alarm clock. It's just no way to start the day. However, when a Learjet is waiting to take you to Steven Spielberg's office, it eases the sting considerably.

I make it to the airport at the appointed hour and there, to my considerable delight, is an actual Learjet waiting just for me. I love airports and airplanes! Took my first flight at two weeks old and I've enjoyed it ever since. In this moment, I'm incredibly psyched. This promises to be another remarkable day in my life.

I board the jet and take the first of the six seats. The pilot is kind enough to leave the cabin door open (it is 1982, after all). I can see right through the cockpit windows without having to move from my incredibly comfy chair. I ease back and wait for the show to begin. The takeoff is smooth and soon we are soaring just above the clouds. It's always amazing to see the sea of clouds, so soft, serene and endless. It seems such a beautiful place to stroll, but I decide to remain in my seat just the same. We're flying to Burbank, then riding to Warner Studios where Spielberg and his sprawling office await. But it turns out we're not going to Burbank, at least not yet. First, we'll stop in Monterey to pick up some additional passengers.

As we near the Monterey area, a most unsettling sight appears through the pilot's windshield. The usual soft white carpet of clouds is now punctuated with a cluster of mountain tops. As we descend through the bright white layer and the visibility shrinks to zero, I can't help thinking that mountain tops usually have mountains underneath them. In this case, I'm really hoping I'm wrong.

Fortunately, the pilot missed every one of them and landed cleanly on the Monterey runway. He taxied a bit, then came to rest on a vacant section of tarmac. Nearly vacant that is, because just as the plane slowed to a stop a big black limo pulls up right off the left wing. The doors open and out pops Ray Kassar (CEO), Skip Paul (Chief Legal Counsel) and Lyle Rains (Coin-Op Game Engineer). Apparently, Lyle is doing the arcade version of

E.T., and I'll bet he's getting more than 5 weeks, too! OK, I didn't really think this last part. After all, this is only 40 hours into the project so I'm not bitter yet. As they file onto the plane, I hear Skip say to Ray, "What? They couldn't get the Hawker?!" He sounds disappointed, but this is hard for me to imagine. They take their seats and away we go. The takeoff is carefree since mountains aren't nearly so scary on this side of the clouds.

We fly for a while more; Ray and Skip are chatting a bit, but Lyle and I are silent. The time of the presentation is approaching, which means the tension and the focus is building. We land in Burbank airport and once again, just as the plane comes to a halt another limo pulls up alongside. "It's just like in the movies," I think to myself, which makes sense since we're going to meet Steven Spielberg at Warner Studios. This is so cool. I can hardly believe it's a workday… but it is, which makes it even cooler. I'm loving this.

We get in the limo, and it's a remarkably well-appointed vehicle. In addition to the plush seating accommodations, there is a phone, a TV, a small fridge, even a sink. Skip reaches over and pushes the lever to watch the water stream out, but nothing happens. The amazing thing was the look on his face. He says, "Do you believe it, the water doesn't even work." OMG! He's serious. This guy just got off a private jet into a waiting limo and he's actually annoyed that the water isn't running in the car's sink. I realize we're from different worlds, and much as I'd like to belong, I'm not really a part of his. I'm always interested to get a glimpse into other people's perspectives. Not always relieved, but definitely interested.

The guard waves us through the gate at Warner and we proceed along the lot until we arrive at the office. We go in and pleasantries are exchanged all around. Now it's presentation time and Lyle goes first, which gives me a little time to chill. My thoughts begin to drift. Spielberg's office is small… for a luxury apartment. It's nice to be back here again. A calm settles in… but not for long. "Wait a minute," I think to myself, "why am I here?"

It occurs to me I don't have an answer. I realize it's because I said "yes" of course, but why did Ray call me directly? That's never happened before. This has all been so exciting, I forgot how odd it was. Atari is big on secret culture and back channel communications, there is always something going on you don't know about. Here's what I didn't know:

I was not the first one Ray called about doing the E.T. game. His first call was to George Kiss, my grandboss (or boss's boss). George is the head of engineering for the Atari home game system, and he told Ray what any sane and knowledgeable person in that situation would: You cannot do a game in 5 weeks. It's simply not enough time.

Most CEOs do not like "no" as an answer. It rarely contributes to shipping product and making money. So, after being told by the head of development it couldn't happen, Ray still thought it was worthwhile to make one more call. I had apparently built enough of a reputation or made enough of an impression that he believed I might come through when others couldn't. Or it might have to do with the time Ray saw my personal notebook and asked to peruse it. I lent it to him, and it came back through interoffice mail a few days later with a note attached. "Thank you, Howard. You are a Renaissance man." This is the nicest thing anyone can say to me.

This was all very flattering and, as I think about it now, rather creepy. I told Ray it absolutely would happen, right after my grandboss told him it couldn't. That's what I didn't know, and I'm glad I didn't. Talk about undermining relationships.

Suddenly, the question, "Howard, what have you got for us?" pierces my reverie and brings me back to the moment. Now it's my turn and I begin my presentation...

The last time I presented something to Spielberg was early June, about a month and a half ago. We met at the Consumer Electronics Show in Chicago and I had *the tape*. I was nearing completion on the Raiders of the Lost Ark game, my second project for Atari and my first for Spielberg. Atari needed a way to demonstrate the game for Spielberg in Chicago. I could have simply played it for him, but I thought it would be better to make a demo tape that could serve other promotional purposes as well. The execs agreed and sent me to a video recording studio to make the demo.

Have you ever done something absolutely perfectly? At exactly the right time? I did. Just once. At that studio.

They sat me down, put a mic on me, hooked up the console to a recorder and I played and narrated the entire game flawlessly. That had never happened in any of my demos, before or since. It was a magical moment. A one-take wonder. We added a few special effects, created a master and that was it. By the way, the total running time was 12 minutes and 27 seconds. If it takes you longer than this to play all the way through Raiders, you probably didn't make the game.

From the time I left that studio in Sunnyvale until this meeting in Chicago, the tape never left my side. There was NO WAY I was going to miss seeing Spielberg's reaction.

Full disclosure: I'm a huge film buff, and Steven Spielberg is a hero of mine. I love his work, from "Duel" on. I think Raiders of the Lost Ark is a masterwork and I was honored to be a part of it in this way. But I'm not just meeting my hero, I'm working with/for him. It's one thing to meet your idol, it's another to have them evaluate your work. It's another still when they evaluate your work which is a derivative of their work. This is huge for me... as long as he likes it.

For a serious creative person, a lot of self-image (and mental well-being) is on the line at a time like this. I was confident but very nervous. I'm one of the top video game creators of my time, but what I really want to be is a film director.

Finally, the moment came. There I was, up in the crow's nest of the enormous Atari show booth with a TV and a tape deck and Steven Spielberg. I inserted the tape and hit PLAY. Spielberg watched it thoroughly and intently. He didn't move at all for the entire 12 minutes and 27 seconds. I know because I watched him thoroughly and intently for the entire 12 minutes and 27 seconds. At the end he thought for a bit, soaking it in. Then he looked up at me and said, "That's really great, Howard. It feels just like a movie!" My inner world exploded with joy. Steven Spielberg thinks the demo tape of my game for his movie feels like a movie. Yeah BABY!

That was one of the greatest moments of my life... but that was then and this is now. I finish laying out the design for the E.T. game and Spielberg thinks for a bit, soaking it in. Then he looks up at me and says, "Couldn't you do something more like Pac-Man?"

My inner world collapses.

Something more like Pac-Man?!?! One of the most innovative film directors of all time wants me to make a knock off? My impulse is to say: "Gee Steven, couldn't you do something more like 'The Day the Earth Stood Still'?"

Fortunately, my brain kicks in microseconds before my mouth engages. Get a grip, Howard. This is Steven Spielberg, and he obviously likes Pac-Man. My father's words came to me in this moment, he was fond of saying "Get your head out of your ass, wipe the shit from your eyes and focus!" Ah, the memories.

All this takes a fraction of a second in my head. Then I regroup and take another tack entirely. "Steven, E.T. is amazing and we need something special to go with it. This is an innovative game for an innovative movie." I believe this is true, but I'm also aware of another fundamental truth: The game I'm proposing is one I might possibly *finish* in 5 weeks, which is a critical component of success in the overall delivery process.

That's why I need to defend this design with everything I've got. I'd rather not fall back on this explanation because I'd rather not come off as desperate, but I will if I must. It harkens back to one of the great linguistic contributions of computer science: Doability (noun, the quality of being able to be done. From the modern English; Do + Ability). Ask any software engineer about the prospect for a task or design, and the answer will invariably revolve around the word "doable." I'm confident this design has sufficient doability to be worth pursuing. This is distinct from another contribution: Bogosity (noun, the quality of being bogus, a mangle-ization of Bogus). Bogosity and doability are independent properties. In other words, creating a game in five weeks can have significant doability and still represent a high level of bogosity on the face of it. In other other words, the possibility of doing something doesn't make it a good idea. I believe this paragraph stands as proof of that.

[NOTE to the Non-Nerd: Many people do not consider nerds to be facile linguists or communicators. Be advised: New-Word construction and deployment is an essential part of the nerd repertoire. To be clear, I'm talking specifically about techie nerds or geeks. Word nerds and/or grammar police are beyond the scope of this text.]

After a few moments of breath-holding, Steven relents on the Pac-Man proffer and accepts my assertion that the design is appropriate to the task at hand (the punishment fits the crime). As he does, I realize my design is now approved. The first major milestone is achieved, my inner world is resurrected, and (though I'm not 100% sure about this) there seems to be a faint emanation coming from Steven's chest, a sort of reddish glow. I have a theory about this…

But this is no time for theory. There are hard facts to face:

- An accepted design only opens the door to begin continuous crunch mode. It is truly the gift that keeps on taking.
- Tomorrow is day 4 of the 35.5 days allotted for the task, 10% of my schedule is already gone.
- I still have to make it through a Learjet ride home before I'm anywhere near dinner! (OK, not all the facts are hard)

The design is now set and approved. It's implementation time. There's nothing to it but to do it!

And as the golden light of late afternoon kisses the flats and backlots of Warner studios, the Atari delegation boards the waiting limousine and sets off for the airport.

CHAPTER 3
ENDINGS & BEGINNINGS

IT IS MY DUST-INY

It is not by Design that I find myself here in this desert. At the tender age of two weeks I took my first plane ride, leaving Colorado behind for suburban New Jersey where I'd spend the next 18 years cultivating a life destined for upper-middle class greatness. At least that was the plan at the age of two weeks. 56 years later I'm protecting my face from sand and swirling garbage in Alamogordo, New Mexico. This was never where I was heading, but it's certainly where I've arrived.

How did I get here? Decades of choices, steps and missteps, muddled intentions, dreams pursued, detours taken, opportunities missed and a long parade of abrupt turns. As is so frequently the case in life, the path which led me here was as unforeseeable as it was inevitable. I hate the word "inevitable." I don't believe in fate. Fate is simply where we run aground when we stop paddling on the river of life. It's the opposite of a plan. It's random chance, a circumstance free of intention. If you want to change your "fate" in life, just start paddling. Chance vs Destiny is an either-oar situation.

But the fact remains I *am* here today, which validates the inevitability. And it was unforeseeable largely because I never really learned to paddle. Consequently, I've spent much of my life adrift... albeit a rather focused drift. This is due to the two main factors guiding my life: boredom and ignorance. I need to do something because I'm bored, and I don't know what that is because I'm ignorant. So, I meander along from one endeavor to the next, finishing some, abandoning others. Doing anything and everything, in hopes of discovering the "right" one.

Here's a brief history of Howard:

As a preschooler I was very focused on what I was going to be, because I knew I didn't want to be what I was, a child. I wanted to be a firefighter, a scientist, a cavalry officer, an astronaut, a ballet dancer, a pilot, a police officer and seemingly always a performer. In fact, I used to do performances in the living room while the adults watched TV, at least I tried to.

I'd yell "Watch this!" and they'd say, "Wait for the commercials." A reasonable response except for one thing, I loved watching commercials. They were compact stories that fit my limited attention span. I'd clamor for attention during the show, which I'd never get. Then during the commercials they'd all say, "OK, let's see it!" At which point I'd sit down to watch the commercials. Lather, rinse, repeat. It was a grand recipe for frustration, both theirs and mine. Little did I know frustration is a crucial component of video games. I was being expertly groomed.

I always liked words and talking. Words are the secret code to getting what you want, and I was always trying to break the code. It launched a lifelong fascination with language, but that would only last the rest of my life. I even interviewed with the National Security Agency to be a code maker/breaker. Spoiler alert: That didn't work out.

I also like beginnings. Fresh material is exciting to me. I like the steep part of the learning curve, but the big thing about new beginnings is opportunity. From a very young age, I found myself constantly wondering: Am I finally arriving where I'm supposed to be? Will this bring real satisfaction? I kept checking because I knew I wasn't there yet. But I was ever hopeful, seeing each next turn as another spin of the wheel, another chance to win the karmic lottery. When it doesn't happen, I'm disappointed. But soon I return to wondering where else to look. How will I find it? I know it's out there somewhere. I'm an optimist, damn it!

It was a comfortable life growing up in suburban New Jersey, but I knew it wasn't my life. Every day I'd go to the mailbox hoping something inside would launch me on an exciting adventure. I knew my path had to lead elsewhere. I was right about that. I also knew I would never get married. I wasn't quite on target with that one.

Once institutional schooling began, my goals became more immediate:

In elementary school: Get into middle school.

In middle school: Get into high school.

In high school: Get out of high school.

Most of my life up to and through high school was largely about doing nothing... but still graduating. Admittedly not lofty goals, but goals nonetheless. And by June of 1975 they were achieved.

Then came college. My high school record was solid enough to get me rejected from most of my targets. I was finally accepted by Tulane University in New Orleans. I never believed high school would contribute much to my ongoing life, but I did think college might. I decided it was time to stop screwing around. Reversing years of philosophical commitment, I decided to apply myself to academics. This would require a different manner of goal setting. I chose to major in Economics as an entering freshman. So, how did that go?

As freshman in college: This can't be it. Maybe I'll finish here and go to the University of Chicago for a Ph.D. in Economics.

By mid-sophomore year: This can't be it. Maybe I'll add a math major, some theater and a computer course.

By late-senior year: This can't be it, but this computer stuff is cool. Maybe I'll get a quick Master's in Computer Engineering.

In graduate school: This isn't bad, but it can't last. Maybe I'll get a job in computers.

At Hewlett-Packard (my first "career" job): This should be it, but it isn't. Maybe I'll find something fun to do in the meantime, en route to... what exactly?

You may notice a pattern here. I'm not sure what I want to do, but I know it's not what I'm doing. This leads invariably to the unsettling thought of having to leave if I want to be happy, but not knowing where to go. This keeps happening. I don't like it, but what can I do? I just press on, seeking my next karmic lottery ticket.

OK, maybe that's a bit too brief. Here's a less brief (but hopefully still succinct) history:

Early on I realized the world people want me to accept is different from the world I perceive. Too frequently I'd be told how things were when I could plainly see they weren't. I'd ask people to explain this discrepancy, but alas, kids don't get real explanations. This was the root of much consternation on my part.

And it didn't take me long to catch on to the disempowerment of childhood, which I never much cared for. This was highlighted for me one day during my sixth year. I asked my mother to pick out my clothes like usual. But recently she'd started saying, "You're six years old, you can do it yourself." Which would be fine, if it weren't for the fact that my choices invariably came up wanting. She'd take one look and say, "That's no good. Do you really think this matches? Ugghhhh." Then she would end up "resetting" me after all. Nothing new there. (In fairness I should add that, to this day, no one who cares about me trusts me to dress myself.)

Later that same day I asked if I could join a friend on some adventure and she told me, "You can't do that, you're only five years old." This offended my five-and-a-half-year-old sensibilities. "How come when you want me to do it, I'm 6 years old but when you don't want me to do it, I'm only 5?"

This was the first of many times I would challenge my mother's reasoning in rule making. No satisfying answer was ever forthcoming. Logic could not pierce my mother's parental-authority-armor. If it made her squirm inside the helmet at times, she didn't show it. The typical response in situations like this was always some variant of "Because I'm your mother." I'm sure she found it frustrating to be questioned. It frustrated me too. Appeals to the court of dad validated my arguments, but to no avail. The fix was in.

The arbitrary and nonsensical nature of so many parental edicts was disturbing to me. I soon realized there would be no relief until I reached the great promised land of adulthood. Because one thing was absolutely clear to me: Adults get to do whatever they want.

Another source of pain from my childhood was boredom. Mostly because boredom was the absence of distractions from looking inward, rarely a pleasant experience for me in my youth. It took me a long time to realize there are always interesting things to ponder, rendering boredom largely unnecessary.

Balancing some of these early observations are two things: My optimism and my sense of humor. I always believe a better situation lies ahead if we can just find the path to reach it. This is an endless source of passion and inspiration for my future. And for my present, I'm usually able to find some fun or joy in whatever is happening now. I'm also rather fond of ironic circumstances, which seem to abound around me. Given this menu, I can always spot a ray of amusement, even in the most dour of circumstances. People don't always join me in this pursuit, because I don't always use as much discretion as I might. The truth is, I'm constantly bombarded with (potentially) amusing ideas and it's hard to resist putting them out there. I see it as product testing. Others may describe it in less kind terms. The fact remains my sense of humor is intact, even if there's no tact in my sense of humor. That's just my nature.

There's also my nurture. Making others laugh is a high-value skill in my family. My father taught me the power of humor, and my mother taught me the need for it. They showed me how using it wisely can lighten things without damaging the severity or sincerity of the situation, although I've come to see not everyone believes in this perspective. Humor can also reset us from a difficult place, revealing a path forward when none is apparent. This is something I use quite a bit in my work and my life. It was certainly handy in my teens, as I tended to take things rather intensely then. Without humor, things could get pretty ugly.

I also believe every problem has at least one solution, if I'm wise enough to see it. As a child and teen, boredom was a problem. My solution was playing games. Games were my go-to anti-boredom device. Be it card games, board games, games of skill or games of chance, I was fascinated by them, interested in them, paid attention to them and thought about them. I loved modifying games. I enjoyed making up new rules and new ways of playing, anything to increase the fun of the game. Can a lame game become fun? Can a fun game be a little more fun? Maximizing fun was always my goal. It was rare that a game was so well-conceived that I couldn't think of any way to improve it, or at least enjoy experimenting on it. But the idea of making a living with games never entered my consciousness.

School was tough on me for two reasons: It rarely held my attention, and my name is Howard (which is a problem if you don't pay attention). "Howard" sounds just like "How are" when I'm daydreaming. Too many questions begin with "How are …" Imagine waking up in the middle of a question from the teacher. My heart racing, my head scrambling to recall that from which I'd drifted, desperately trying to spare myself the agonizing humiliation of being caught once again. The question finishes, the moment hangs, and just before I'm forced to admit my lapse… someone else answers or the teacher simply continues. Ah, sweet reprieve, but recovery from the adrenaline rush is not instant. Welcome to my K-12 experience, repeatedly waking up from daydreams into daymares. A seemingly endless series of mini traumas. I'm reminded of the old proverb: When you add a little to a little, you get a lot.

Shortly after turning fifteen I entered high school. For some reason, this led me to seriously contemplate the rest of my life, and the picture I saw stretching out before me was overwhelmingly underwhelming. I had a brief existential crisis: "Is this trip really worth the fare?" The ultimate upshot of which was: "Isn't it up to me to make it worthwhile?" (apparently my internal reflections manifest in rhetorical questions) In that moment I committed to leading an interesting life. I wasn't sure what that meant or how I'd do it, but I made myself a promise that if I was going to bother to live an entire life, I was going to do everything I could to make it worth the time and effort this invariably takes.

So, the first thing I did on my bold new adventure was watch as much TV and play as much poker as I possibly could. OK, this isn't the kind of 'next step' you might expect from a life-altering internal revelation, and I agree completely. I'd created a good intention, but it would take three more years to begin implementation. Let's just say: Though the seed was planted, it still needed time to germinate.

And germinate it did, just in time to replant it in the fertile soil of my college career. I had come to the decision that college mattered, something I never felt about any prior school experience. I didn't know what I ultimately wanted to do, but college seemed worth doing. So, I picked things that sounded reasonable and did the hell out of them, just in case they turned out to be useful.

How did college go? Pretty well, considering I had no previous experience (or desire for) being a good student, which can be an issue in college.

I attended Tulane University in New Orleans and got educated to a significant Degree, but one of the greatest lessons of my entire college experience was not part of their curriculum. It was a lagniappe (as they say down in The Big Easy), provided by none other than Dick Cavett, noted American wit & TV personality (who wasn't even a Tulane faculty member).

He was moderating a discussion at a Tulane speaker symposium. The topic was anthropology, and it featured two prominent anthropologists of the time. One was Margaret Mead, a great speaker and joyful spirit. The other, whose name escapes me, was a bit pretentious. Let's call him the Stuffed Shirt. I came mostly to see Cavett, I admired his quick mind and subtle style. Anthropology was only of mild interest to me. Little did I know it would reenter my life a mere thirty-six years later, at the epicenter of an actual dig in pursuit of my yet-to-be-created creation.

The discussion was lively as Margaret, Cavett and the Stuffed Shirt went at it. At one point, Cavett made a witticism that didn't sit well with the Shirt, whose response demonstrated how "grace" was no part of his stuffing: "Mr. Cavett, if you're not capable of taking this seriously, why are you here?" Without missing a beat, Cavett looked the Stuffed Shirt right in the collar button and said:

"Please don't mistake my levity for shallowness any more than I mistake your gravity for depth."

BAM! Indelibly etched into my brain. These words spoke to a core issue of my life. It rubs me the wrong way when people try to reduce me with limiting assumptions. It's also fun to zing them when they do, and Cavett's response deftly handles both so beautifully. Humor is an important part of who I am, but it carries a downside at times. People assume if I can joke then I'm unable to grasp the solemnity of a situation. I feel there's room for both, I don't aim to be insensitive, but I don't do somber well either. As you may imagine, I'm not always great at funerals. I've never cared for others dictating how I should feel. When they do, it both saddens and pisses me off. But it's bothered me less since I heard Dick Cavett that day. In one shining moment, he gave me an amusing and healing insight, my favorite kind! I'm still appreciating it.

Thanks to the Tulane University faculty and Dick Cavett, I did well enough in those four years to convince Hewlett-Packard to hire me. They moved me out to Silicon Valley in California, and finally it was time for me to step into the working world and be the grownup I had always wanted to be. At long last, adulthood beckoned.

I was tremendously excited to get to Hewlett-Packard. Finally, I'd taste *real life*, the place I'd always looked forward to and worked so hard to get to. With my computer-assisted passion driving me forward, the wind felt wonderful on my face. And it *was* exciting… for a little while.

But all too soon the joy of computer programming drained away. I thought I'd found my path, my destination, my key to the kingdom. I should be happy, but I wasn't. I still hadn't clued-in to any ultimate direction. Once again, I'm stuck in that place where I don't know what I want, but I know it isn't this. I'm back at sea.

Approaching the end of 1980 at Hewlett Packard, this was truly the autumn of my discontent; The hardest yet. My sadness was deepened by my belief I'd finally found the answer with my joy of computing in college. That was gone now. It turns out the only answer I'd found was that it truly is worse to have loved and lost. This launched the first major depression of my life (if you don't count the majority of high school).

Then, in an odd twist of coincidence, my autumn of discontent was transformed into a winter of resurrection when the prospect of video games raised its head and caught me in its tractor beam. During a random conversation with a coworker, the subject of Atari came up.

Atari? I'd seen their TV commercials. I knew they made video games, but I'd never thought of them as a place of employment. Why would I go into video games?

How could I not? I'd spent my entire young life grooming myself as a video game programmer. After all, I love games in general and I'm a hybrid of inventor, systems analyst and entertainer, the perfect resume for a job no one knew existed a few years earlier. And how did I prepare myself? Entirely without awareness.

Inventing was one of the few things that captured my attention as a child.

I'd pile up my broken toys and game pieces and try to create something new. I rarely did, but I kept sitting there with invention intention just the same. When I'd read, it was usually about inventors and their inventions. What's cooler than inventing things?

Systems. I was always interested in systems, particularly human systems. This was strictly self-defense. I knew there were currents and secret pathways which led to getting what you want or need. I believed there was some way of navigating that elusive world of adults which got you there. I was ever the hungry outsider looking in through the window and drooling.

As for the entertainer in me, I have my parents to thank for that. After all, they raised me to be an Entertainer. Don't get me wrong, they weren't Showbiz parents at all. They just made sure I had plenty of insecurities, a need for approval and no concept of boundaries or limits. The perfect qualifications for winding up… in a desert sandstorm?

I ended up here, but where did the path to here begin? Was it birth? Was it the phone call from Ray Kassar? Beginnings and endings. That's what Atari was all about. Punctuation. Game projects started, and they ended. Each game is a fresh start. I like that. One irksome aspect of life for me is summed up thusly:

There is no punctuation in life, period!

Brilliant moments, devastating losses, fun, pain, joy, tears, nothing stops the wheels from turning. Birds keep flying, clocks keep ticking, life never stops. I always longed for real definitive moments in life, and Atari did a great approximation. But I digress.

My first day at Atari was a new beginning for sure, but also the end of a long search. My search for destiny and direction. A search for identity and uncovering true desires.

Before we talk about arriving at the design and execution of the E.T. video game, let's talk about what it was like arriving at Atari. This was the beginning of a very different way to think about what constitutes work and professional environments. Eye opening doesn't begin to describe it. Mind-blowing is more like it. Such a memorable time, and another one of those amazing days in my life…

WELCOME TO ATARI

It's January 12, 1981, the second Monday of the year. My first day at Atari starts typically enough for a new job, but nothing could prepare me for the realities of the world I was entering.

As a video game developer, the Atari universe comes down to two buildings on either side of Borregas Avenue in Sunnyvale, California, known internally by their addresses: 1272 and 1265.

1272 is Engineering where all the games are made and 1265 is Atari world headquarters, which houses management, marketing, sales, HR and basically everything else that isn't manufacturing or engineering. But 1265 had the game room, a mini arcade with all the Atari coin-op games (set for unlimited free play), which is basically the only reason someone like me might head over to 1265. But in 1272, we have the hot tub, the gym, and the cafeteria. Besides, as a game developer, more time is spent playing prototypes of new games than the released games so there wasn't much need to head over to 1265. In fact, for people based at 1272, the frequency of trips to 1265 was a measure of your rank in the company. The higher up the ladder you climbed, the more time you spent "crossing the street."

All my interviews were in 1272, or in a trailer parked behind the building which functioned as a temporary appendage of the building. Atari facilities were nearly always in a "we're about to" state, so any office is considered transitional/temporary pending a move. Most moves are permanent for at least six weeks before discussions about moving resume. So, when I report for my first day, I go to the only place I know, upstairs in 1272. The first thing they do is send me to 1265 for orientation. The HR reps tell me about benefits and such, but nothing about what life at Atari is actually like. Honestly, I'm not sure they know. After watching the introductory video and filling out some perfunctory paperwork, I head back to 1272 for whatever is next.

[NOTE to the Non-Nerd: Atari has three places for a programmer to work; Home Computer, Coin-op games and Console games. Coin-op games are the ones in arcades and you pay each time you play. Console games are played at home on a box hooked up to your TV. They are purchased once for a much higher price and then you can play forever. Atari's Console, the box upon which home video games are played, is more commonly known as the VCS or the 2600.]

I sit down with Dennis Koble, now my manager, and he asks me a question that will define my entire experience for the next several years. He says: "We have VCS or Home Computer. Which do you want to work on?" In a few months, with the gift of hindsight, this choice would be painfully obvious for an ambitious guy like me. But this is now, and I have no reason to prefer either, such is the plight of the newbie. So I do what I usually do in situations like this: I answer the question with a question. "What's the dirtiest, ugliest, most primitive system you have here?" I hate moving from higher capability to lower capability systems, so I figure if I start at the bottom, I'll save myself anguish later. In life, most of our choices come down to one of two approaches: maximizing pleasure or minimizing pain. Since I cannot yet grasp the pleasure potential of these options, I choose the latter criteria.

"OK, that's the VCS." Dennis assures me. And quick as that, my fate is sealed. I'll be making console games.

Among the innumerable implications of my answer, the very first is seating location. As a VCS programmer, I'll be somewhere in the main quadrangle. 1272 is partitioned into several territories. Downstairs is Coin-Op engineering, the cafeteria, hot tub and gym. Upstairs is VCS engineering and the Home Computer Division.

The VCS department occupies the top floor of one wing of the U-shaped building. Its main feature is a central hallway that cuts the quadrangle into an inner rectangle and an outer ring. The interior contains restrooms and the labs where development systems live. There are no windows here. Actual programming, debugging and tuning work happens in the labs, where everyone else can see what you're doing. The outer ring is for offices. They do have windows. People sit and think and pencil out code in the office, then go into the lab where they test the code on a development system and figure out how to fix it. Offices also serve as an incubator for shenanigans. The hallway itself serves as a test track for all manner of racing or path-oriented games we might want to explore, as well as hosting the execution of many shenanigans, which are an integral part of life at Atari.

Dennis gets up from his desk and motions for me to follow. He takes me to an office in the far corner of the floor. It looks like a storage closet for desks since the room is full of them, leaving precious little space for a person. I'm the third human assigned apparently, "You'll be in here with Tod & Rob." At the time it meant nothing to me, but I would soon learn that the phrase "Tod & Rob" means quite a bit around the quadrangle. I enjoy contemplating the unknowing beginnings to major parts of my life. Tod Frye and Rob Zdybel, two amazing characters who remain friends to this day. They would become formative people in my life. Yet in this moment, standing among the sea of desks, they remain unallocated neurons in my brain for a few hours more since neither is here now. I'm confident they exist because two of the desktops near the door have things on them. Grateful for the opportunity, I clutter a third with my things and continue the journey with Dennis. As we leave the office I can't help wondering: Will Tod & Rob notice my things and contemplate my existence?

Dennis leads me around the quadrangle and I'm taking note of the faces as we go. Some are familiar from my interviews and some are new. Dennis is quick with introductions. He turns left through a doorway and as I make the turn to follow, my eyes pop. It's one of the labs where games are made. All around the room, big televisions sit next to big black boxes on the upper shelf of the workbenches. Keyboards and printouts sit on the desktop. Interesting things are happening on every one of the TV screens. Some simple, some elaborate, but each with colors and motion and beeps and pings, it is hands down the coolest moment I've ever enjoyed in a workplace. In front of each station, seated on a tall stool, is an Atari game engineer. In this moment I realize these people now have an additional designation in my mind: colleagues. It strikes me how wonderful it is to be here. This is the exact opposite of boredom!

One thing in my life had always been painfully clear: I don't know what I want to do. I can't imagine a place I'd fit in. I've only ever known "this isn't it." But as I stand here in this lab, I realize "right here, right now" is precisely where I want to be. For the first time in my life, I feel I've arrived. We walk around the lab and Dennis is introducing me to the developers but I'm not really taking it in. I'm overwhelmed with pure joy.

After the lab, he shows me where supplies are kept, and who are the key admins to which I might turn in times of need. Finally, he hands me the standard starting package of documentation and a few floppy discs to use for my yet-to-be-assigned project. He assures me I'll have an assignment very soon and encourages me to explore coworkers and documentation in the meantime. This whole introductory process is much less formal than my previous job at Hewlett-Packard. HP has a much more elaborate orientation process. I don't have much basis for comparison, but this works fine for me. I'm not one to stand on formality, unless I'm jumping up and down trying to crush it.

Returning to my newly assigned desk, I reflect a bit on my day so far. Seeing all the screens in the development lab, it occurs to me these are video games soon to be released. Millions of people out there are eagerly anticipating this stuff. They can't wait to get their hands on it. I'm at the epicenter. This is broadcast media and I'm in the "studio." The entertainer in me perks up, this is my chance to show them what I've got. I hope I've got something to show.

Sitting here in the office is a bit of a fishbowl experience. People walk by consistently. Some are curious about the new arrival and some are not, but my eye is drawn to each of them. I don't want to shut the door and hide, but manual reading is something best done without distraction, so I head downstairs to the cafeteria, which is empty since it isn't mealtime. The traffic is far less frequent and I'm feasting on new data.

After a while, Jim (one of the programmers) happens by and stops to chat, which is nice. While we're talking, I can't help noticing there's something attached to one of the belt loops on his jeans. A second glance confirms it's an alligator clip, and judging by the discoloration on the tip, one that's been used recently as a roach clip.

[NOTE to the Non-Head: A Roach Clip is a mechanical device used to hold the last bits of a marijuana cigarette (colloquially known as a "roach") so as not to burn one's fingers while extracting maximum value. Alligator Clips are spring-loaded metal clips. Their elongated array of interlocking teeth resembles the snout of an alligator. They are occasionally mistaken for Crocodile Clips.]

Assuming he forgot about it, I point and say, "Hey, I think your roach clip is showing." I don't want an oversight on his part to cause any inconvenience for him. Jim looks down at the clip, smiles at me in a way that suggests how cute it is that the new guy thinks this could be an issue, "Oh, that's not a problem."

From the start it was abundantly clear marijuana was a real presence at Atari. Even during the interviews, people were feeling me out on this topic, which was not a concern. I was perfectly happy to participate in this aspect of the culture on occasion. But the fact it's so blatant is a new concept entirely. I'm beginning to realize I may need to recalibrate my sense of workplace propriety.

Jim asks me how I like it so far? I tell him I'm really digging it and looking forward to being here. Then he says, "Yeah, it's pretty good, but it isn't what it used to be. This place used to be amazing." Admittedly I don't know much yet, nonetheless this statement feels odd to me. I cannot imagine a better work environment. But I know sometimes people lose sight of the forest for the trees and fail to notice how cool something is because it's become normal. I chalk it up to bourgeois entitlement, as I have no intention of ever taking this environment for granted. In my experience so far, Atari is beyond amazing, and no comment is going to change that.

After reading a while more, I decide to head back up to VCS and continue my socialization. I walk around the hallway, look around the labs and ask a few questions. Strike up a brief chat here and there about my background and what game they were currently developing. On one pass down the hallway, this happened: Someone (presumably a game engineer) is walking in the opposite direction. As he passes, I notice he's making a series of sounds I cannot decipher. This is pre-cellphone, so he is not talking to anyone in particular and I don't recognize the language at all. Stupefied, my eyes and ears are compelled to follow him down the hallway until he turns the corner and disappears. Apparently, seeing someone stupefied by this display is not that unusual, because another hallway passerby notices my quizzical expression and offers this unsolicited but welcome explanation: "Oh, he was raised with a twin and they created their own language. He thinks out loud in it." Thankful for the explanation, a smile comes over my face. For no reason I can articulate, this makes perfect sense to me and I feel more a part of everything for the knowledge. I am so glad to be here right now.

I return to my office and peer out the huge plate glass window. The early nightfall of winter assures me the end of my first day in the new world is fast approaching. I'm getting a few more pages of manual under my belt when Tod breezes into the office, the door closing shut in his wake. He produces a small plastic baggie whose contents appear green with purple sparkly streaks. He takes note of me and, with the ultimate in nonchalance, begins the most unforgettable exchange I'll ever have on my first day of a new job: "I'm going to smoke a joint now. If you don't want to be around this, you should leave."

I appreciate Tod's courtesy in offering me the choice and assure him leaving will not be necessary. In fact, I'd prefer to stay and join in, which is fine with him. Recalling my interview experience, I am prepared today with a joint of my own, freshly rolled this morning. Isn't that part of everyone's pre-work ritual on the first day of a new job? I reach into a pocket and produce my contribution, offering it to Tod. He gives me a look which could be either disinterest or disdain (but no hint of surprise whatsoever) and says, "No offense, but I'm going to smoke some *real* stuff here."

Apparently, my new office mate is a pot snob. As the newbie, who am I to argue? I'm here to learn. He opens the baggie, releasing a much sweeter aroma than I anticipated. He then proceeds to roll up a joint rather adroitly and we smoke it. Soon it becomes clear to me I'm not dealing with a snob after all. Tod is a connoisseur. His stash was far better than mine. I realize I'm going to have to up my game on a variety of levels in this new job. After spending some time chatting most enjoyably with Tod, it was time to call it a day.

On the way home, I sift through my experience of the last ten hours. It's super exciting to be at the core of a magical engine, generating new kinds of entertainment that millions of people around the world are hotly anticipating. I'll get to see the games, play them, influence them and even create them on a daily basis.

I know that, for the first time in my life, I don't want to be somewhere else, or someone else, or do anything else. Finally, my foreseeable future is perfect just as it is. This is what makes Atari so remarkable. Something very special is going on here, and now I'm part of it, which is glorious!

It is also clear Atari will totally reset my concept (and expectations) of life in the workplace, much to the chagrin of a long series of future managers.

At the same time, Alligator-clip Jim's words are still ringing in my ears, "this place isn't what it used to be." Imagine showing up at your ideal job, and hearing people complaining how it's gone downhill. How it was sooo great before, but now look at it. I'm thinking, "Yeah, look at this. It's a dream come true. WTF are you talking about!?!?"

It turns out I'd landed in the middle of a huge cultural transition. One with ominous ramifications for Atari, the game industry and the world of technology. What I'm not quite grasping yet is when people say, "It used to be better," what they mean is, "I don't like where management is heading." A new wind is blowing, less sympathetic to developer needs. I am too close, too new and my eyes are too wide to see it. But I will come to understand the truth of it and, impossible as it seems now, I'll be singing the same tune soon enough.

I've stepped into a life-warp, wrapped in a reality quake. Atari will totally redefine my sense of who I am, where I'm going and what I need. Or perhaps Atari will simply enable me to finally hear the answers I've known all along. Either way, this is the first of over a thousand Atari days in a row, but I don't know that yet. In fact, at this point I have no concept whatsoever of the extraordinary adventure ahead of me. I don't yet realize that the promise I made to myself at fifteen years old has just started coming true.

I do know this: I can't wait to come back tomorrow morning! I need that in my life. There weren't any days like this at Hewlett-Packard, but I'm feeling one right now and loving it.

And I know one other thing: I've finally found a home.

FIRST ASSIGNMENT - YARS' REVENGE

After two days of manual reading, I get my first assignment: Convert the coin-op game "Star Castle" to the Atari VCS game console.

[NOTE to the Non-Nerd: Copying existing coin-op games was a common approach for VCS development and a sensible marketing strategy. If you like this game at the arcade, why not play it at home anytime you want? But there's an issue with this strategy: Coin-op game technology improves with each new game while home console tech (like the VCS) stays the same. Since the VCS has already been out for a few years, the coin-op tech is already way ahead of the VCS, making it harder to deliver decent conversions. The way to improve home console games is discovering/exploiting previously unrecognized capabilities... or designing games that use it more cleverly. This fact will prove most fortunate for me... eventually.]

OK, here we go!

The key to successful project planning is knowing what you're trying to do. Logically you might assume my goal here is to bring Star Castle to the VCS, but you'd be mistaken. This is my first video game for Atari, consequently my design goals are:

1. To Make a Splash – I want my debut to be a real contribution, something that will establish me as a player in the world of video game makers.
2. To Create a Sensory Experience – I want it to be a distinctive, eye/ear catching extravaganza that cannot be ignored.
3. To Break New Ground – I don't want to iterate on existing material, I want to create something fresh and innovative.

Am I asking a lot? Sure, but why aim low? This is what I'll strive for and the results will fall where they may. Now, let's look at Star Castle as a candidate and see how it fares with these three goals. After playing the game for a while and travelling to arcades to observe expert players, I come to the following conclusions:

1) Splash? Star Castle is a clever game. It was created by Tim Skelly for Cinematronics, and it was a remarkable technical achievement at the time. It has some innovative mechanics, but the particulars of this game will be a nightmare to recreate. I'm still new to the Atari system, but I've learned enough to see this game is going to *suck* on the VCS! Will it be a contribution? More like a charity case.

2) Sensory Experience? The on-screen motion is interesting, but the visual focal point is stuck in the center of the screen. Black and white line graphics are not known for stunning visuals. In fact, all the color in this game comes from a plastic overlay. Not the eye candy I'm looking for. The sounds feel a bit limited and somewhat monotonic. Star Castle is kinetic but not dynamic. I need to do better.

3) New Ground? Let's face it, when you're doing a knock-off, innovation is not the thrust of the work. It didn't take me long to realize that coin-op conversions are the opposite of what I want to do.

My best shot is to make an action game I will enjoy playing, and it's clear that Star Castle isn't going to do it. This is unacceptable.

I see the Atari game development world as a club, a club to which I very much want to belong. My first game is essentially my application for club membership. I'm not just making a game; I'm creating a calling card. It's tremendously important to earn my place amongst the Atari game developers, and converting Star Castle is not going to do the trick, so…

Clearly my first task is changing the assignment, thus creating room for me to do something better.

As a veteran of a week or so, I go to Dennis and tell him straight up that this game will suck on the VCS. Then I share my plan to take some of the basic mechanics and tweak them into something that will play much better on our console. OK, it wasn't that simple. I made a whole presentation out of it, including sample screens and renderings I'd created using graph paper and everything. Fortunately, Dennis was receptive to the idea and said, "Go for it!"

I can hardly believe my luck. He's letting me do it! This would never have happened a year later, it's a good thing this is now.

The transition from coin-op conversion to free-range game is accomplished. Now I'm unrestricted and my options are wide open...

So, what do I do now?

I was forced to do something people in technical production rarely consider: Fall back on my education. I have degrees in Computer Engineering, Mathematics, Economics and Theater. The Computers and Math got me the job, but the Econ and Theater will prove key to my success as a game maker. Supplemented by my innate flair for drama, of course.

One thing theater taught me is to evaluate the work through the eyes of the audience. What do I want the player to experience?

I want my first game to be a frenetic symphony of color and motion, combined with twitch action. The player is in constant danger, if you stop moving, you're dead. It must be dynamic, compelling and irresistible if possible, as well as visually stunning! As people walk by, they are forced to stop and say, "Wow, what is that?"

The film buff in me also contributes. It should be as much fun to watch as it is to play, with game action drawing the eye all over the screen. The sound should enhance the experience as well. I want to create a soundscape that dictates mood and builds tension.

These are all important design criteria for my first action game. It has to make a splash. I need to innovate. I'm barely starting, but I'm trying to expand the idea of what's possible on this machine.

Like I said, why aim low?

But at some point, I still must answer the question: How exactly will I accomplish this? Here's where my economics background comes in handy. After all, economics is the science of allocating scarce resources, and the VCS is a desperately limited resource for programming. All I hear about the VCS is how restrictive it is. I'll approach game making as an economist, which means maximizing impression while minimizing cost and effort.

[NOTE to the Hard-Core Nerd: Was the VCS limited? We had 4K of ROM for code and only 128 bytes of RAM for game state. Seriously. Individual data structures today are bigger than our entire product. Console games today are more than a million times bigger. They're definitely better, but are they a million times better?]

OK. Great theoretical approach. But what will I *do*?

The truth is: I don't know. I'm not sure yet, but I'm confident this is a journey of discovery. I'll just start banging away, doing the few things I know must be done, and see what possibilities present themselves along the way.

Two things I've got going for me: 1) my obstinate nature, and 2) I frequently do things in atypical ways. That's just me. I compulsively flex my perspective, searching endlessly for unseen possibilities. Whenever things seem too limiting, I start looking for loopholes and escape valves, ways to engage things that previous explorers might have missed or ignored. And I never stop, which annoys some people. But occasionally something useful pops up, which is a most welcome thing in a creative environment.

This is why Atari is so perfect for me. Finally, I'm in an environment where being offbeat and different is an asset rather than a liability. I'm nestled into a group of misfits, and I fit perfectly.

By the way, what makes a good creative environment? How do we plan to be creative and host new ideas? This question has haunted creators for a very long time. Fortunately, I'll be attempting to answer this in one of the most creative incubators ever assembled, and I'm just getting warmed up.

CREATIVE ENVIRONMENTS: PRODUCTIVE VS. WASTED TIME

The sweeping majesty of huge dust clouds billowing over the desert floor is an interesting counterpoint to the steadfast determination of the big yellow machines. Unmolested by blustery gusts, the mechanical behemoths stand resolute in their ongoing effort to dig and dump. The power of nature vs the product of the toolmakers. The chaos of the storm is mirrored in the hive-like undertakings of the attendees, except for one thing: The storm is throwing garbage around randomly, the human activity seems far more coherent.

The movie people are scrambling around looking for their next shot, but they move in squads with each member towing their special accessories. The food people buzz around their trucks, preparing to feed the needs of the crowd. The news people are getting their equipment checked, covering any openings where sand may intrude and trying to stabilize reflectors against the gusting winds. Lights flash on and off. Boom mics wiggle in the wind. Interviews are constantly starting and finishing. And all this is happening amid a sea of hopeful enthusiasts, running from place to place in every direction with their souvenirs and heirlooms clenched tightly in hand. All are braving the elements for a chance to see history happen before their eyes and lenses. A cavalcade of seemingly random activity, yet every movement has a purpose, each step has a goal.

All these factions doing their thing, interacting with each other. All this energy and commotion cohering into a harmonious productive mass. This is what I see here in the desert. It's organized mayhem.

But is it productive? There is no guarantee anything of note will emerge from the depths. What if nothing does? All this work and effort and attention, it could all be for nothing. If no games pop out of the hole, is this experience meaningful? Does it yield anything of value or wind up an elaborate waste of time and energy?

This reminds me of a Q&A session I did at a Classic Gaming convention in Texas. After sharing some storied shenanigans, one audience member asked me this: Wasn't a lot of time wasted at Atari?

It's a great question, and a tricky one to answer.

Organized mayhem is a wonderful description of life in Atari's VCS department. Differing agendas with a common goal: pushing for something exciting to emerge with no guarantee of success.

What does it mean to spend time productively when you're innovating? You might spend 10 months diligently creating a crappy product, finishing on time and under budget. Or you could spend 2 months goofing off and end up with a killer idea which produces an excellent product just 7 months later. Which way is better? Which is wasting time? How do you measure it? Which model would you use to build your project plan? If you can only measure results yielded in retrospect, that's a tough problem for planning.

And what will you do during the 2 months of goofing off to maximize your chances of coming up with an idea that justifies the time? There's still no way to guarantee it. On the other hand, it's easy to spend a predictable amount of time on a mediocre product that goes nowhere and loses money. And if you do, is that a total waste of time from start to finish? You'd be better off financially not doing it in the first place. But don't forget to balance the money results with the experiential aspect. Did we learn something here that will make our next product dramatically more successful? It can be tricky figuring out what to do. What you don't want to do is spend 2 months goofing off and then take 10 months delivering the original mediocre product late! That much we know for sure.

To my view, creativity is taking unrelated things (or ideas) and putting them together in a fresh way (possibly counterintuitively), creating a new capability or opportunity. How can I make this happen?

Organized mayhem is what you're looking for in a creative environment. It's what I'd call a necessary-but-not-sufficient condition. There are also key roles critical to the mix. There must be Stimulators to generate the mayhem, and Inspiration Lightning Rods (ILRs) who can capture worthwhile moments amidst the chaos. That's crucial, because if you can't catch the lightning bolts when they do strike, all the time and effort creating them is truly wasted. Both these roles must be present. At Atari we had Stimulators, ILRs, people who were both and some who were neither. But the neither's were good at taking the creative output and delivering it well.

That's fine for what we need, but there are equally important things to be avoided in a creative environment. Two that come immediately to mind are conflict and corporate expectations. Both of which sprang increasingly from our interactions with Marketing as time went on. I'll get deeply into the issues between engineering and marketing later in this book, I promise. My good friend Bob Saenz, who is an award-winning screenwriter with many scripts produced, hates this very much. He does not hate the conflict between engineering and marketing, because that generates compelling scenes for his movies, but he hates when I say, "I'll tell you about it later." I know I shouldn't do that, and thanks to Bob, I'm doing it far less here than in prior drafts. But I want to finish the creative environment part first. For now, let me just say, "Sorry, Bob."

As recompense, I will say this: Marketing likes things reliable, predictable and neatly organized. Engineering wants things unpredictable and chaotically organized, to enhance the opportunity for a breakthrough inspiration that we can then deliver in an organized fashion. These approaches work for each in their respective corners, but when the bell rings and the two mix? Atari founder Nolan Bushnell, widely known as the "father of video games," puts it beautifully: "[Engineering needs] an environment with little discipline and yet with clearly stated goals. In general, that's in conflict with a corporate form."

I've worked in creative environments where people think conflict is necessary for inspiration. I disagree. I believe conflict and competition are distractions. I believe organized chaos with low conflict, high comradery, and enough courage to keep pushing limits and busting through barriers… that's the most potent recipe.

At Atari, we would do things in the hopes of becoming inspired. Trying new things, trying old things in new ways, looking for a different point of view. The fresh perspective.

Perspective is fun to play with, particularly when it shows us how we limit ourselves. Is the glass half empty or half full? Perhaps the glass is too large. And what's in the glass? Am I thirsty? Shifting my perspective opens new doors, creating new sources of light. Perspective is about solving problems. I believe there are no hard or easy problems, there are simply helpful and unhelpful perspectives. We solve problems by adopting more useful perspectives for the situation. Here is an example: Logic puzzles.

Remember a logic puzzle you couldn't solve? It seemed like a really tough puzzle. Then you learned the solution and suddenly it was an easy puzzle. But the puzzle never changed, you did! You incorporated a new perspective and that changed everything.

Locking ourselves into limited perspectives shrinks the world. When we are free to adopt different points of view, more solutions become available to us. We become more effective problem solvers when we work less on our problems and more on ourselves.

Which of my life issues might be simplified with new perspectives? How can I learn to change my perspective more easily? More productively? In my therapy practice I help people answer these questions all the time, but that's another book entirely.

As Atari game engineers, our job is to generate new ideas that create fun. How do you do that? What does a productive creative environment look like? We'd just do wacky things, looking for fresh perspectives and different points of view.

After all, games were our business. Playing games was our research. Inventing games happened frequently and spontaneously. There was little concept of goofing off because any time-wasting activity could potentially be a game concept, which was work. As long as you wasted time actively and not passively, you were contributing to the research effort. But how do you know which you're doing? If you lapse into pure goof-off, the clock is still ticking. Rob Fulop summed it up nicely: "The distinction between work and play becomes so blurred, that if you aren't really disciplined, you might never get any work done."

We played bocce lemons, which means throwing lemons down the hallway trying to get closest to the target object, which was whatever someone threw down the hallway that was not a lemon. This answered the critical question: What is it like to play a rolling physics game with a ball that is not actually spherical? It also raised a question: Should we get mirrors mounted at the corners of hallways so as not to bombard innocent passersby with flying lemons?

Moving into new offices happened pretty regularly at Atari. I had five different offices in four years. When you got to a new office it was discovery time, because you never knew what it used to be. Office space happened as it needed to and it seemed facilities planning was not a top priority.

One time, we moved into an area that used to be some manner of industrial lab. This led to a new pastime. They had super-pressurized air outlets which were still active. We gathered various food stuffs and took turns jamming them into the nozzle before popping the valve. The sudden burst of air either vaporized the material or launched it across the room, slamming into the opposite wall. Bananas made for the most agreeable results. Important note, if peas are not cooked first, they become dangerous projectiles. OK, since our game console did not have any olfactory interface, this one might have been an actual waste of time. It was a lot of fun though.

There was a period when we used to play demolition derby in the parking lot with remote-control model cars. That was a fairly expensive indulgence, though it did serve to release some pent-up stresses and frustrations of creative expectation. The hot tub came in handy for this sort of relief as well, especially when used in creative ways.

We used our tools, our toys, our facilities and ourselves in odd and unorthodox ways. We were innovators, working at the State-of-the-Art. The thing about innovators is they tend to have boundary issues. That's how they avoid being victimized by limited "rut" thinking. However, when boundary busters cut loose, there can be collateral damage to innocent rule abiding bystanders. When you break containment, things can get out of hand suddenly and quickly.

Like at company parties. There were three kinds of parties at Atari. There were the big formal inter-departmental corporate events which were rather staid. Then there were the small intra-departmental affairs which were decidedly more bacchanalian. They'd start off nicely enough around a large table or two in a local pub or restaurant, but they could wind up in any of many directions. Attempting to keep the crew orderly (or even reasonable) was a fool's mission, as many local business owners will attest. In fact, even keeping it indoors proved challenging at times. One year a game of parking-lot-chicken broke out. This is where slighter employees sit on the shoulders of more substantial employees and try to "decouple" opposing teams. If the bushes and ground look soft enough (because everyone is drunk and/or stoned enough), this can seem like a pretty good idea. Despite extravagant tipping and the fact that no ambulance ever needed to make an appearance, most venues opt never to host a 2nd Atari party. I don't blame them. The third kind of party was the "weekly" Friday afternoon decompress, which was a less outrageous hybrid of the first two. It happened most weeks.

In addition to the parties, there were offsites and boondoggles of all kinds. Working the booth at the Consumer Electronics Show (CES), attending focus groups, consumer testing events and various conventions. As time went on, there would be meetings with high profile partners and license prospects.

Hmmm, when I list it all together like this, it sounds like the only thing we didn't do is work. But these things happened sparingly throughout the year. Most of the time, we were at our keyboards, trying to realize the value of the stimulation and inspiration we garnered from all these divertissements.

The entirety of this was funded by the deluge of dollars flooding in from selling successful video games to the entire world. The payoff of a productive creative environment is clear… but the process remains elusive. As I said earlier, video game design is about seeing things in interesting and fun new ways. But how do I plan fresh ideas and creative breakthroughs?

Our approach was doing wacky things in the hope of stimulating insight. At Atari I was always looking to freshen my point of view by engaging mundane tasks in an unusual manner. And not just on my dev system. For instance…

During most of my Atari tenure I sported both beard and moustache. One day, just to change things up, I decided to go clean shaven for a while. Whenever I contemplate a change like this, I pause to consider the unique opportunities the transition may provide. In this case, I started to think about what I might do with my beard that ordinarily I would never do, because this was surely the time to do it. I decided it would be fun to shave half of it off and go into work for a couple of days like that before finishing the job. So, one morning I shaved off half of my beard and the opposite half of my mustache, sort of checkerboarding my face.

The first thing I noticed was how few of my colleagues noticed, which irked me a bit. I more or less had to get in peoples' faces before they'd notice mine. But that's really just a testament to how ensconced people were in their own projects/world. More interesting to me were the research opportunities, like the department meeting which happened to come up that afternoon.

After a little practice, I was able cover half my face with either hand. If I used my left hand I had only a beard, with my right hand only a mustache. The department meeting was a presentation by some outside person, so I went early enough to ensure a front row seat before the guest speaker arrived. I brought my right hand up to my face, covering the beard part and rested my elbow on the table and waited for the meeting to begin.

The guy showed up and began his presentation. I'd hold my position and keep eye contact with the speaker as much as possible. After a few minutes, during a moment when he turned to the board, I switched hands. I continued to switch each time the opportunity presented itself. At first it went totally unnoticed, but after a few rounds he would pause an extra second to look at me before continuing. Next time his look became a bit more quizzical. The next time I did it, he turned back around from the board, looked at me, gave up any pretense of continuing and said, "What is going on with you?" I pulled both hands from my face, shrugged my shoulders and said, "What do you mean?" As he took in my patchwork barber-ism, the most quizzical expression came over his face, as if to say, "Who would do this?!?" It was priceless.

Of course, had he asked the question aloud, most people in the room would simply have said, "Oh, Howard would. Or Tod." That's just how it was in the VCS department. His face made the whole experiment on my face worthwhile.

Many interesting predicaments and happenstances proceed from chasing creative inspiration. One of the main challenges at Atari was thinking up ways to stimulate the flow of creative juices.

One of the more intrepid adventurers in this regard was Tod Frye, who gave us one of the all-time examples of the unpredictable ramifications of coaxing the muse. It's one of Tod's more memorable moments, known by insiders as the Sprinkler Lobotomy…

It starts the way great stories do, with contempt for limits and boundaries. We were trailblazing this new field of techno-tainment, and all good pioneers share one common trait: the irresistible impulse to bump up against rules and standards, find the soft spots and bust through. Remember: Innovators are people with boundary issues. Be it technology, standards of propriety, authority in general or even chemical tolerance, we tested the limits! That's just who we were. This time, Tod's target of opportunity was gravity.

It may help to understand that Tod used to work in construction, so he had a rather intimate relationship with structures, both inside and out. Not many people explore a building like Tod, much to the relief of first responders everywhere.

In our latest building, there are hallways of different sizes. Some are narrow enough that Tod can jump up in the air, put one foot on either wall and steady himself about twenty inches off the ground. This is cool... for a couple of minutes. But for Tod, this isn't just a new capacity, it's an opportunity for expansion of concept. As Tod puts it in the Once Upon ATARI documentary series, "We're engineers. What we do is develop a new system and explore its capabilities... and then we develop new features." True to his philosophy, Tod explores what he can do with this newfound way of engaging a corridor.

First, he graduates to inching his way up the walls vertically. It gets to the point where he can get a good five feet off the ground before running into the ceiling tiles, but this will not suffice. Having mastered the vertical climb, he starts exploring lateral movement. Initially he can make short clunky moves, gaining a few inches at a time. But eventually Tod discovers that by shifting his weight from side to side, he can begin to rhythmically walk down the hallway, well above the floor. Soon he can run, and at a pretty good clip. This is not only impressive, it's impactful. By that I mean the banging of Tod's feet against the walls was deafening. BOOM! BOOM! BOOM! It is never a secret when Tod is doing his thing. The jiggling of wall hangings and the loud report of feet hitting wall panels speaks unambiguously. If you're in your office and hear Tod coming, you know to stay put until the Doppler effect tells you the coast is clear. Otherwise you can wind up with a sneaker print on your chest or face.

Before long, Tod's wall-walking becomes a normal part of an Atari day. They vary in duration depending on the number and location of open doors. An open door is the bane of wall-walking. Whenever Tod comes to an open door, he simply "dismounts" by jumping down to the ground and reverting to a more pedestrian mode of transportation (e.g. walking).

One day, Tod is testing how high "up" can be. He is running particularly close to the ceiling when he encounters an open door. As he releases his feet he jerks upward slightly, nudging his scalp into one of the sprinkler heads hanging from the ceiling. The scrape catches Tod by surprise as the sharp protruding edges cut deeply into his skin. Tod falls to the floor, dazed and bleeding profusely.

The ensuing crash and screams of agony are not Tod's usual wall-walking sound effects, so several people rush to the scene. After some back and forth it is decided Tod must go to the hospital for evaluation and treatment. Two people volunteer to take him and away they go. During some after-discussion, it is further noted that Tod has left some hair (and a bit of blood) on the sprinkler head. We decide the hair should remain in place, to serve both as a souvenir of the event and a stark reminder of the hazards endemic to our trade.

Tod and the others arrive at the emergency room. Naturally, the admission people want to know what happened. Tod begins to explain how the unanticipated fire sprinkler had cut short his latest jaunt, as well as his still-bleeding scalp. All Atari engineers are used to hearing Tod relate his exploits, with the concordant meta-levels, self-observations and philosophical digressions. But this time Tod is actually giving a cogent and concise recounting of the incident. Naturally, the nurses don't believe him.

In an effort to build credibility, Tod launches into the origin and evolution of wall-walking. Somehow the backstory didn't make the hospital staff any more inclined to buy it. Suspecting Tod was speaking from delirium or concussion, they kept seeking affirmation from the other engineers. But the others backed up his story 100% and so, despite the incredulousness of the admitting desk, Tod was ushered in for treatment. In the final analysis, I suspect the blood was more convincing than the story.

Tod assures the others he'll be fine, and they head back to Atari, but not before making sure Tod calls Dave Staugas and tells him Tod's at the hospital. Dave is a maker of games and taker of photos, but most importantly right now, he is Tod's commute buddy. They both live in Berkeley and work in Silicon Valley, so they carpool together for two hours a day. Dave is concerned to hear the news, but it's not so much shock as curiosity. When you know Tod, you realize there's little point in reacting to dramatic pronouncements until you get the whole story. This isn't a lack of caring for Tod, it's to avoid burnout because with Tod, some bit of drama is never far away. Interestingly, Dave will eventually come to work with us at Atari, despite witnessing firsthand the dangers inherent in the job.

In the ensuing hours Tod's wound is cleaned, treated and dressed for success. The area beneath the wound, his brain, is assessed and reviewed multiple times as well. His repeated recounting of the accident is as consistent as it is unbelievable. I'm not sure they believe Tod was running down a hallway at shoulder height, but I think they believe they aren't going to get another version of the story. Silicon Valley emergency rooms see plenty of programmers, but never quite for this reason. Ultimately, they clear his mental status, finish up the paperwork and release Tod to Dave for the ride home.

The whole episode is captured beautifully by a phrase which not only speaks the absurd literal truth of the matter, but also communicates the metaphorical experience of so many video game developers:

> Reason for Visit: Patient injured while climbing the walls.

And that's the Sprinkler Lobotomy story. A classic piece of Atari history. It demonstrates how creative people find aggressive and sometimes dangerous ways to further the scope of their expertise and experience in search of elusive breakthroughs. Perhaps that's why denizens of Silicon Valley sometimes refer to State-of-the-Art as the Bleeding Edge.

We were entertainment engineers navigating a creative environment in search of insight and discovery. What does "creative" look like? How much spark is needed for quality product? These are questions we endeavored to answer. An amorphous goal to be sure. It was trial and error, and we had to show our work.

It wasn't like we didn't have any structure or process, sometimes we even sought auxiliary support for creative inspiration. One important research activity at Atari was known as the MRB. MRBs were initially held in the women's room on the second floor of the engineering building (1272). After a while though, it was deemed poor form to see men walking so nonchalantly into (and out of) the women's room, so MRBs were moved to other locations. Despite the way this sounds, it really wasn't untoward. The women's room had a lounge just inside the door. One had to pass through there to arrive at the bathroom facilities proper, so it wasn't a matter of decency. There was, however, the consideration that forcing anyone to walk through an MRB on their way to the bathroom could constitute a discourtesy (if not a felony), so relocating was really in everyone's best interest.

On a semi-regular basis an announcement would come, "MRB in so-and-so's office" or "MRB in the upper addendum" which meant an area on the roof of the building if it wasn't raining. Unlike department meetings, MRBs were not mandatory. There were, however, various levels of membership. There were charter members, semi-regulars, occasional attendees and some who saw fewer than one. MRB is the acronym for Marijuana Review Board.

The announcement would be made, either by intercom or "Paul Revere" style, the collective would gather, and combustibles would be oxidized. Usually a great deal of productive discussion ensued. Game concepts were spawned and/or elaborated, bugs were fixed and technical challenges were resolved during MRBs. Also, many strains of Humboldt county's finest were assessed for quality along several key dimensions, because one thing you learn at Atari is that product testing is a very important part of the creative process.

So, when that audience member in Texas asked me, "Wasn't a lot of time wasted at Atari?" I simply told him the truth as I saw it: "Time wasn't wasted at Atari. *Engineers* were wasted at Atari!"

There has always been a lot of speculation about drug consumption at Atari. I'd like to settle this speculation once and for all by assuring you that lots of drugs were consumed at Atari. Not universally, there were abstainers as well as indulgers, but a variety of substances were consumed in a variety of ways by a variety of people from a variety of departments. No drug was consumed by everyone and no one consumed every drug, but every drug was consumed by someone at some point. Of course, those drugs were merely chemicals.

Did they boost or hinder productivity? Creativity? These are topics for debate. But here's one thing that's undebatable:

The *real* drug at Atari, the one everyone in product development chased obsessively, was getting your game released. Let me tell you about that high. Walking into a store and seeing your game on the shelves, or better yet, playing on the demo systems. Seeing ads for your game on national TV. Your work, climbing the Billboard sales charts. If you are a person who makes things, this is tremendously validating. Validation is something everyone in the entertainment business seeks, and Atari was definitely in the entertainment business. It masqueraded as a technology company, but it was totally an entertainment business. I loved that about Atari.

THE DEPARTMENT MEETING

Was your 25th birthday memorable? Or, if you're not there yet, are you planning something major? Mine started out interesting and progressed to absurdly unforgettable. It's the Friday after the Tuesday on which I received "the call". If you told me on Monday my birthday would fall 10% into a development schedule that didn't exist yet, I'd have asked to share some of what you are obviously smoking. And that would have been a nice birthday gift. Little did I know, a much better present was already in the works.

The day is already primed by the fact I'm waking up a mere twelve hours after my return flight on the private jet from the presentation in Steven Spielberg's office. My design is accepted, and all systems are go. I have

33 days left to deliver the game and no plans for today other than work. Revelry be damned! I have all of September to celebrate my birthday if I want, so off to work I go.

Upon arriving, I am greeted with a subtle twist, too subtle for my hyper-focused tunnel-visioned project-maniacal state-of-mind. As I walk the halls to my office, I can't help noticing all my co-workers seem to be wearing orange, the exact same shade of orange. We do not have uniforms at Atari, so this is odd. In fact, it's the same shade of orange you find on the Yars' Revenge product packaging, which it turns out is not a coincidence. On closer inspection I realize they are t-shirts, bearing a design that looks eerily familiar. The image on the shirts is a caricature of me. In it, I'm a Yar, complete with wings, antennae and even my trademark sandals. I was also cracking a bullwhip (ala my Raiders of the Lost Ark game). Underneath the image (in bold lettering) is "**HOW-WEIRD!**" A nickname I had acquired through every fault of my own.

The entire department is wearing them. My girlfriend had orchestrated the whole thing. She got one of the graphic/animation designers to draw up the caricature, then got them printed on t-shirts and arrived early enough to distribute them before I got there. It's amazing and I totally love it!

As luck would have it, there is a department meeting set for this morning. We occasionally have meetings for the entire VCS department. I enjoy making jokes or cracking wise at these meetings, as opposed to other times around work when I like to tell jokes or make wisecracks. One typical example would be the time we were discussing the state of the development labs. We have three labs; A, B and C. For some reason the door to Lab A was missing and no one knew why. I immediately pointed out that this would be easily solved, in fact there are animals specifically bred and trained for this purpose. When everyone gave me the "What are you talking about, Howard?" look, I explained how we simply need to get a Lab-A-door retriever. Anyone who's worked with me is familiar with this syndrome. I'm not saying they're happy about it, just familiar.

MARILYN CHURCHILL & HSW AT THE DEPARTMENT MEETING.

PHOTO BY: DAVE STAUGAS

So, people are dribbling into the meeting as usual, and everyone is wearing their HOW-WEIRD t-shirt. I have to say, as department meetings go, coolest look ever.

Eventually everyone takes a seat and the meeting comes to order. George, the department head, announces that I am doing the E.T. game. And the mood of the room shifts. It is not lost on the crowd that I had just finished putting the final touches on Raiders of the Lost Ark. The echo of my bullwhip still reverberates down the empty hallways. Yes, my bullwhip. Allow me to explain…

While working on my second game, Raiders of the Lost Ark, I decided it would be fun to get into character. I went out and found a nicely weathered fedora and a very authentic 10-foot bullwhip. I practiced with the whip until I could crack it without injuring myself. When I'd hit a good one it was incredibly loud, like a gunshot.

Occasionally I would take breaks from programming and roam the halls with my hat and whip. Just like you could tell when Tod was walking the walls, you knew when I was away from my dev station. Now and then I'd spot a familiar marketing rep, walk up behind them and startle them with a good one. The managers in our department endured regular uncomfortable moments explaining odd phenomena to uninitiated visitors. Making sense of the loud cracking sounds echoing down the hallways was one of them. Accounting for that odd smell of burning rope was another frequent duty. The VCS department was an interesting place to manage.

One day I was strolling around "in character" and decided to pop into Lab B to see what was up. Lo and behold, here is a news crew doing a "filler" story for tonight's newscast. This happened occasionally, since Atari was always good for some quality B-roll. As soon as they saw the whip, the camera turned toward me and the interviewer approached, mic in hand, "Is that a real whip?"

"Yes it is."

"What's it for?"

Holding it up conspicuously for the camera, I said the only thing I could think of in the moment, "This is for R&D. Research and Discipline." I'm guessing this led to an interesting discussion in their production meeting that evening.

Back here in our department meeting, a rumble of grumbles is emerging from several corners, they are variations on the theme: "How come Howard gets to do all the big titles?"

This makes me uncomfortable. I don't like to feel the chill of my colleagues' cold stares (unless I've just told a painful pun). Also, there are many Big Titles around and I've only done one, so I don't quite get the "all the big titles" aspect of it. But hey, I'm an equal opportunity kind of guy. So…

I stand up and say, "Hey everybody, E.T. is due September 1st (less than 5 weeks). If anyone wants it, just say so. You can have it."

crickets

It's so quiet, you can hear a bit drop. An aura of disbelief comes over people as 'September 1st' sinks in, but not a single murmur of protest. I sit back down, the meeting goes on as usual and I never hear another complaint about me doing E.T. At least not until the mid-1990's with the proliferation of the internet. Even then, no one has ever said to me "I wish I'd done E.T."

There is, however, a major shift in the narrative. It's no longer "Howard gets to do all the big titles." Now it's "OMG, Howard is out of his mind!"

This was never about being selected to do E.T. This was about being *willing* to do it. Or as some put it, being dumb enough to do it. Either way, I was ready to put myself on the line and take the chance. It's a pretty ballsy thing to do. I'm risking my professional reputation on a very high-profile project in a ridiculously tight squeeze. Perhaps I didn't fully grasp the implications. But there was no doubt in my mind this was something I needed to do, and it was a cinch no one else was willing to attempt it. Honestly, if I could go back, knowing what I know now, I'd still do it. I'd make some changes of course, but more to the negotiations than the design. Don't worry, we'll get to that soon enough. Sorry again, Bob.

The drama of the moment subsides, and the meeting continues. After issues noted and announcements made, the most monochromatic meeting in VCS history finishes up and all the orange t-shirts return to their respective offices and development stations. And that is that. Which is not to say the topic never comes up again.

As the whole thing simmers over time, there are programmers who share concerns with me about doing a 5-week game. Originally people were keen to say it can't be done and it's stupid to try. But once I accepted the task, they started to consider the ramifications. What if it *could* be done? What will this do to future schedules and expectations? What if the suits start thinking two months is a generous schedule? This is not a ridiculous line of thought at Atari, and there are some who consider it unwise to even attempt this sort of thing.

But one thing was certain: No one else would touch this... and that was exactly what I wanted. I was looking for a mountain to climb, a huge challenge to validate myself and that's when the E.T. project showed up. Be careful what you wish for.

I stride back to my office and dive in. Scrambling to cobble some code and throw some test screens up during the afternoon. As 5pm approaches I'm called to the enormous graphics room under some pretext, only to discover a lovely party assembled in my honor, complete with a cake bearing the caricature from the t-shirt design. I even get to make a speech, and everyone is forced to listen. What a great birthday! I love this job and

I love these people.

I'm eating cake and reveling in the festivities, it's all good. Then there comes a point when I realize I'm only able to relax and enjoy this because I know I'll be here the entire weekend, relatively undisturbed. It's clear to

PHOTO BY:
DAVE STAUGAS

me if I'm going to pull this off, I need to find a way to separate myself from all this. My workplace, so full of vibrancy and stimulation, the manna for my soul, is being reduced to a distraction by my limited schedule. It occurs to me the things I love most about Atari must be put away for this project. In the middle of this beautiful party, it is a sad and painful moment.

Early next week, a development system is set up in my home.

CHAPTER 4
HERE COMES THE CRASH

The New Mexican desert seems to be getting hotter. Partly because the storm has finally relented and partly because it's noon and the day is heating up like deserts do. The sandstorm punished, but it also cooled. At least we are no longer being assaulted by flying debris, however short-lived that relief may be.

As the huge claw buckets keep delivering older news from greater depths, I'm sensing a steadily growing sense of fear in a few of the crowd members. Some of Alamogordo's city leaders worry that mercury gas and other toxins might issue forth from the fermented garbage of yesteryear as we violate the protective layers formed by decades of dumping. Me, not so much. But a cloud of negativity and bad karma? This seems a real possibility. I went to Atari not for games but for microprocessors and culture. The culture was amazing yet there were people complaining it used to be so much better. I couldn't imagine it being better, but I did see it get worse. In fact, I saw it all tumble down. The toxic culture clashes at Atari are legendary. I saw the foundation crumble beneath my feet as the greatest parade in history proved too much for the bridge.

Atari was the Kleenex of video games. Nobody called it their home game system, or their VCS, it was simply "my Atari". Wanna play Atari? Did you get anything new for your Atari? In the late '70s Video games burst on to the scene like a tidal wave. Atari was the fastest growing company in the history of American business. How does something like that just disappear?

In 1983, the video game world seemed to spontaneously combust. Evaporating at the height of glory, leaving only questions. What happened to Atari? What happened to video games? How does a multibillion-dollar industry suddenly go poof, as if by magic? What kind of magic has this much power?

Peering out over the horizon, I recall the first atom bomb test occurred near

Alamogordo. How apropos. The story of the great video game crash of the 1980s is a lot like an Atom Bomb. Did you know it takes two detonations to create a nuclear mushroom cloud? First you have to set off a sizable charge inside the bomb which creates an implosion. This generates enough energy to set off the main explosion, the big boom. The crash of the video game industry is also the result of two detonations. Atari's philosophies and business practices generate the initial implosion, clobbering the company. Then, as the consequences of Atari's actions (and inactions) reverberate and spread, they ultimately set off the big BOOOOM, blowing away the industry.

The crash was something no one saw coming, or perhaps some people did, but most were caught by surprise. That's the way it is with crashes. If enough people saw it coming, they would take corrective action and it wouldn't happen. But it happened, and as my E.T. game became the face of the crash it magnified the ignominy significantly.

Over time I've developed my own ideas about the symptoms, causes and warning signs of the crash; a multitude of tributaries combined to form a mighty river of inevitability which ultimately burst through the dam of denial, flooding the video game valley and washing out the town. Hmmm, suddenly I'm feeling a bit thirsty.

The incredible phenomenon of video games was just like this morning's sandstorm. It came out of nowhere, inserted itself everywhere, turned our world upside down, then suddenly vanished. How was this possible?

MAKING HISTORY BY DESTROYING AN INDUSTRY

What caused the Video Game Crash of the early 1980s? That's easy.

Atari's E.T. video game killed the industry.

Everybody knows that.

In the movie, Atari: Game Over, Nolan Bushnell says, "A simple answer that is clear and precise will always have more power in the world than a complex one that is true." I believe this is the reason many people credit the E.T. game (and me by association) with destroying the multibillion-dollar video game industry of the early 1980s. And for some excellent reasons.

Atari paid way too much for the license. They made millions more cartridges than they sold. And don't forget, it was the worst video game ever made. It ruined everything. Right?

After all, the fastest growing company in history became the fastest falling company in history. Tides turned. Fortunes changed. The sudden destruction of this prolific industry was tragic for many people. Decades of news reports have taught us: When telling the story of a tragedy, don't just give the big picture. To increase impact, personalize it by telling the story through the eyes of someone with skin in the game. Every tragedy needs a face, and this would be no different. Atari's E.T. video game became the face of the fall of video games, and I became the butt behind that face.

The legend stood; the E.T. game destroyed the industry. It's a simple and clear explanation. That's what many people still believe, but those people didn't work at Atari in the early '80s. So, I ask you…

Was E.T. the cause of the crash?

In over 100 interviews I've answered this question many times.

The E.T. video game did not cause the crash, but rather it was a symptom of the thinking and actions that led to the crash. And as is frequently the case, the symptoms are easier to identify than the root causes and dynamics.

Think about it: Let's say Atari lost $50 million on E.T., which they did not. Is $50 million enough to topple a multibillion-dollar industry? That seems unlikely.

But consider this: One of the biggest properties in the video game world was given the shortest development time, by a factor of five. Does that sound like good planning?

If you want a simple explanation that is actually true, I can give it to you. Here is the simplest statement of why the video game market crashed:

It was the first product life cycle.

There, now you know. This was the first time a home video game console had achieved significant market penetration, becoming a mainstay for millions of families. Mistakes were made and lessons were learned. Naturally, it was future game console makers who rode those lessons to success on a road paved with Atari's mistakes. But that's the way it is for trailblazers. As the first born of two firstborns, I can relate. My mother was always quick to remind me: "You make all your mistakes on the first one." Thanks, mom.

Though true, "first product life cycle" is a gross oversimplification, and I don't want to be gross. So I shall now endeavor to under-simplify it for you, for a very good reason: The story of the demise of Atari and the early video game industry isn't simple. It's a complex web of characters and circumstances and agendas and motivations and nuanced ramifications. It's a big story. Huge.

And who better to lay this out for you? After all, I hold degrees in Economics, Mathematics, Computer Engineering and Psychology. I also have firsthand experience, having worked in the Atari trenches for years. Add decades of reflection and the result is a variety of insights into the myriad causes of the short-lived death of my beloved business. Many of these same issues ring true today in technology businesses everywhere. I feel they're worth sharing, toward the end of avoiding future crashes (as well as the ensuing retrospectives).

Throughout this book I will occasionally take a detour from the narrative highway and explore the many and varied factors which contributed to the great video game crash. There will be tales of internal conflict, cultural clashes (corporate culture that is), colorful personality traits and some of the shadows residing in the darker corners of our souls. But this is an Atari story, so the fun comes first.

I agree the most eloquent way to tell the bigger story is through the eyes of individuals, the people whose lives were touched and shaped by the events. In this case, that person has got to be Rob Fulop. Rob was one of the first people I met at Atari and has remained one of my favorites. He is an affable sort, a very clever and clear-thinking guy. Rob used to say that the people who could really make it at Atari were the people who wanted to have as much fun as possible and still go to heaven. They could strike that tricky balance between goofing off enough and still getting the work done. Rob was the very model of this philosophy. So, with his kind permission, I'm going to use him to share some key moments in Atari history as well as illustrate some of the underlying dynamics which led to the great crash. I'll do this by sharing my take on his experiences in…

A TALE OF THREE CHRISTMI…

(it's the plural of Christmas)

…as seen through the eyes of young Rob Fulop (and reimagined by me, your host on our journey to Christmi past).

Christmas #1, 1980

It's mid-December 1980 and all the good little boys and girls at Atari are eagerly anticipating their Christmas boni. And no one is feeling the spirit more than young Rob Fulop. He's completed Missile Command for the 2600, and it's a thing of beauty. Everyone knows this game is flying off the shelves. Many months of hard work have resulted in a game which is netting the company tens of millions of dollars. Now it's time for the company to express its appreciation and Rob is listening most attentively.

"What will it be?" he wonders. A check with five fat figures on it? Perhaps the keys to a lovely new car? As the days pass, you can see the visions dancing around his head. Something of substance to commemorate young Rob's dedication and hard work. A token of sincere gratitude. "I've been a good boy, it's time to see what happens to good boys."

At long last the day arrives, and his manager comes strolling down the hall with "the envelopes". And here's the one for Rob.

Rob jiggles it first to see if it rattles, then he holds it up to the light. Finally he peels it open and is treated to a folded sheet of paper, which he unfolds directly. As he absorbs the contents he freezes, absolutely incredulous. Never in his wildest dreams did he dare to imagine his reward could actually be...

A certificate for one free turkey.

Frantically he scours the floor around him for the check he surely missed... he tears the envelope wide, but nothing. That was the precise moment young Rob vowed he would never do another game for Atari, and six months later he left (with his manager who delivered the envelope, and another wonderful Atari programmer) to form Imagic, Inc. Thus, the second great defection from Atari was the product of a fowl-ed bonus attempt. Let this be a lesson to tech-xecutives everywhere.

You see, this turkey didn't happen in a vacuum. Rob had already been sensitized to this kind of treatment by management, largely as a result of the *first* great defection from Atari: Activision, Inc.

Although I did have occasional fun moments with our CEO, it's important to note that relations between Ray Kassar and programmers weren't usually so convivial.

The Activision story is a significant one. Here's how it happened:

One day in the spring of 1979, some programmers came to talk with Ray Kassar. This was not just any group of programmers, they were known collectively as The Fantastic Four. Headed by David Crane, they had produced the lion's share of Atari's best games to date. They pointed out that Atari was making huge profits from their creative work and their compensation was not in line with the value they were delivering. The programmers saw this more like the music industry, where artists get royalties on their work. The people who make the most popular songs (earning the most money for the producers) make the most money from their work. This sounds reasonable to many people, but Atari CEO Ray Kassar is not one of those people.

Ray did not come from entertainment or technology. Ray was executive vice-president of Burlington Industries, a huge textile manufacturer. After decades working his way up the classical corporate ladder, Ray felt he had a very good idea of how a company ought to be run. Spoiler alert: It was not about giving low level employees more money than necessary.

Consequently, Ray was quite clear in expressing his point of view, that being how their contribution to the product was no more than any other line worker assembling the product. In fact, he said they were nothing more than towel designers (something with which Ray did have experience), and if they didn't like it, they could leave because they were easily replaceable. This was pretty much the end of the conversation.

Ray was even quoted later that same year in a major newspaper as saying video game programmers were a bunch of "high-strung prima donnas". I'm not disputing the truth of Ray's observation, but it is also true that high-strung prima donna's get paid a lot more than chorus players, but I digress.

Another thing happened later that year. The programmers in question, having thought this over for a while, came to the conclusion that if Ray believes we're so replaceable, then he should do so. And the Fantastic Four left Atari.

But they didn't just leave. They teamed up with a music industry executive (of course) and launched a new company called Activision. Precious few people recognized it at the time, but this was a moment that signaled the beginning of the end of Atari and constituted the first bricks in the golden highway which led to the destruction (or near mortal wounding) of the early video game industry.

The formation of Activision (and subsequently Imagic) was only a symptom of the incubating disease that would ultimately fell the industry. The real cause is the transition from Nolan Bushnell to Ray Kassar, and what their differing philosophies meant culturally for the company and its employees.

Activision was the first third-party developer in the video game industry. Third-party developers are people who make content for someone else's product. Before this, the rule was: They who make the consoles, they shall make the games. No one else made games for the VCS because the console was so obscure and complex, no one else knew how to do it. But just like you don't have to be a record player manufacturer in order to produce and sell records, anyone could theoretically create and sell a game cartridge for the VCS. It could be done, but no one tried it… until Activision. This changed the entire picture forever. It's a very big deal.

These third-party houses were formed by bitterness on the part of the programmers before they left, and memorialized in bitterness on the part of Atari after they left. It all could have been avoided were it not for the greed of Atari's Warner-spawned management. They were truly penny wise and pound foolish in this case, violating one of the fundamental laws of video games: Greed Kills! Sadly, Atari execs didn't play video games, so they never learned this lesson.

Atari's philosophy was: No suit fits an ex-employee like a lawsuit.

If you can't stop them, sue them. Atari (under Ray Kassar) sported a prodigious legal department. Then they picked up Nixon's former head of security. We suspected they had bugged all our offices, largely because that was the word from one programmer who was sleeping with one of the security guards. We all have our methods of data collection in the information age.

Rob was hired by Atari founder Nolan Bushnell in mid-1978. Nolan founded Atari in 1972, then in 1976 he sold it to Warner Communications for $28 million. As part of the deal he stayed on in a managing/advising capacity. In late '78, Bushnell leaves after being pushed out by Warner, who likely found Nolan a meddling mess as they tried to operate the company "sensibly". No love was lost.

Rob witnessed Nolan's exit from Atari and the rise of Kassar. Rob witnessed firsthand the whole drama between Ray the 'K' and the Fantastic Four. He sees the formation of Activision and all this sets him up to be primed to catch the next train out. The turkey certificate was merely the ticket. It cost me a good friend at work, but it cost Atari so much more.

But it also taught Rob what to expect, and he anticipated a fitting for an upcoming lawsuit of his own.

Now, let's pause a moment to answer a very significant question: Are video game makers rock stars or towel designers?

An odd question? Perhaps, but this was the difference between Nolan Bushnell's upstart Atari and Ray Kassar's corporate Atari. And as time passed, the difference grew louder.

At Nolan's Atari, game makers were held in very high esteem, which means they got to do drugs in their offices and screw around all they wanted as long as when they got a game done it was cool and he could hang out with them occasionally.

At Ray's Atari, game makers occupied a lower station, which means they got to do drugs in their offices and screw around all they wanted as long as they met marketing schedules and he didn't have to deal with them much.

Nolan thought of game makers as unique and important because he knew it took a rare and esoteric combination of talents to conceive a good game and then realize it on the VCS.

Ray thought of game makers as disposable since they dwelt at the bottom of the org chart, where workers are interchangeable.

Game makers loved working for Nolan. We didn't love working for Ray. I hope that's easy to understand.

This first Christmas Story demonstrates how management undervalued their engineers. The turkey is an extreme case, but it illustrates the shift in company mentality from Nolan Bushnell to Ray Kassar. That transition was extremely significant philosophically and culturally. It also had profound implications for the bottom line.

Christmas #2, 1981

There was a certain irony to this Christmas at Atari, because the elusive dream which sent Rob running for venture capital was about to become true for many of us who remained, largely because he and the manager and the other wonderful Atari denizen had left.

It's like the old saying goes: Once burned, twice shy. But when twice burned, it's time to check the grill for issues. The Activision formation merely served to dig Ray's heels in a little deeper, but after the Imagic formation, even Atari executives became concerned about the prospect of any more golden geese flying the coop. By Thanksgiving they presented the first Atari Royalty Plan, which was a surprise. Unfortunately, the plan guaranteed virtually nothing, deferred payouts for years and landed with a pronounced thud, which wasn't a surprise. The fact this wasn't much of a motivator was communicated back to management. Ordinarily, we would expect management to rail against the ungrateful programmers for not appreciating the crumbs cast on the water. But it appeared management actually listened this time, because just before Christmas they came across with some significant bonus money, a big surprise! This made our Christmas at Atari rather jolly. Many programmers became very happy, and we all owed a debt of gratitude… to Rob and pals.

Back in the day, there was no trade show exclusively for video games. There was just the Consumer Electronics Show or CES, and video games were simply the latest gimmick in a sea of click-flash-beep gimmickry. Twice a year, every year, this convention convened at the same times in the same locations, Chicago in June and Las Vegas just after New Year's. Chicago was always fun, but Vegas was amazing. The idea that someone would turn us loose in Vegas (with an expense account) was incredible. You couldn't tell if Vegas made us livelier or if we made Vegas glow, but CES in Vegas was magical. Running around all day analyzing the latest video games and electronic toys, that was our *professional obligation* for the trip. The rest of the time was focused on "allocating the expense account" for fun. Free flowing booze and MRB's abounded. Gambling, shows, great restaurants and our only care was hanging on to the receipts.

Many memorable moments manifested during CES, and one of the best is about Rob Fulop, some Atari execs and a long green-felted craps table. It's two weeks after Christmas of '81. Rob is the guy who created 2600 Missile Command for Atari, then left and created 2600 Demon Attack for Imagic. Both rank near the top of any credible great games list. There were those at Atari who really thought Demon Attack should have been done for Atari. When Rob and the others left to form Imagic, they created a very tough competitor. This move enraged more than a few Atari execs, and none more so than a VP who had befriended Rob and took Rob's exodus as a personal effrontery of the highest order. It's funny how the people who do the least to keep you there rage the hardest when you leave. After Rob, the manager and the wonderful other departed, the vacuum they'd left behind soon filled with an aura of bitterness and ire… on the part of some Atari execs, that is, and this VP in particular.

So, it comes to *pass* (so to speak) (if you speak craps) that several Atari big & medium wigs are gathered around one end of a craps table, and among their number is the aforementioned VP. They are drinking and gambling it up and having a classic Vegas time… when who should sidle up to the other end of the table but young Rob Fulop himself, and the tenor of the evening abruptly shifts. Shooters come (and don't) and things move along as they are wont to do at a dice table. But the banter is taking on a decidedly sharper edge. The testosterone is growing so thick, players are breaking off chunks to tip the cocktail waitresses. Eventually Rob gets the dice, and the Atari execs make a big show of betting against him and they cheer when Rob inevitably craps out. And none cheered louder than the VP.

Now Rob, being the cool customer he is, just sips his drink and bides his time until the dice are finally passed to the VP. After placing a bet, the VP looks up and dares Rob to bet against him. It's childish and absurd, and just the kind of situation Rob cannot resist. He grabs a stack of chips and places a hefty bet. Only instead of betting against the VP, he bets *with* him. This is pure Rob. The VP doesn't know what to do. He hates the idea of Rob profiting, but he can't root against himself. He's angry and conflicted, and trying not to show it. The VP begins rolling the dice... and he is on fire! It's the hottest roll of the evening, and although he's making a bunch of money, Rob's making more. You can see the VP's blood pressure rising with each throw. Never was a crap shooter so irked by their own good fortune. The more the VP tries to hide it the more it spills over. Finally, the VP craps out and everyone at the table cheers for the hot shooter, and no one cheers louder than Rob. The VP is seething. Amidst the chaos and afterglow of the hot shooter, Rob looks over at the VP and says, "You were *so* right."

"About what?"

"You always told me if I stuck with you, I'd make some real money."

It took two directors and a national sales manager to restrain the VP as Rob headed for the cashier's cage with racks of chips in his arms and an impish grin on his face.

Atari didn't build character, it revealed it. There were certainly characters there, but this story is about character traits. Bitterness and pettiness are not too hard to find at any company hosting power struggles between strong personalities. Atari was no exception.

And speaking of personalities, did you know Corporations are defined as "legal persons"? As a therapist, it occurs to me that if a corporation is a person, then it likely has a Corporate Personality. If we accept this idea, it stands to reason they are susceptible to Corporate Personality Disorders, and CPDs are no laughing matter.

When I peruse the pantheon of possible problematic personality predilections, I believe the diagnosis which best accounts for Atari's symptoms is Corporate Narcissism. After all, they believed they were infallible, they expected blind allegiance from everyone, and they were angered by (and deeply despised) anyone who left them or crossed them. They constantly blamed others for missteps, all the while trying to manage their persistent fear of being exposed as the imposters they lacked the insight to fully recognize they were. Corporate narcissism isn't pretty. Does any of this sound familiar?

Christmas #3, 1982

To recap: Rob felt screwed over by Atari in 1980. Then he got some recompense (and very good bonus money) in 1981 to be sure. But now it's Christmas of 1982 and young Rob is on the verge of the best revenge of all... living well.

During the year, Imagic releases many wonderful game titles and is doing well. So well, in fact, that they are about to go public with their stock, or "IPO" as we say in Silicon Valley. This is the dream of every start-up, and it promises to make young Rob, the manager with the envelopes and all their fellow Imagicians very wealthy indeed.

The mood around the company is gleeful, and young Rob is reacquainted with joyful anticipation. You see, in addition to holiday spirit, Rob is once again filled with the reasonable expectation of experiencing the true meaning of Christmas bonus. And the best part is: This year's Santa will not be an Atari exec. This year's Santa is a bevy of Wall Street financiers wearing expensive 3-piece suits, and they always give the same thing. The ghosts of Christmi past will have absolutely no say about what Rob does or doesn't get for his bonus this year. It's the happiest Christmas season ever, and Rob is fully recovered from the ill effects of the 1980 version.

As the days go by and THE DAY draws nigh, the visions dancing around Rob's head are evolving. Images of cars and 5-figure checks are now replaced by houses and endowments, with more zeros than he previously dared to think. And Rob being Rob, he's also anticipating the unique luxury of having money to waste on myriad pranks which Rob dreams up on a regular basis. He's the master of cute gestures and ironic responses to slights. Now, instead of just dreaming them up, he'll be able to do them. In fact, Rob already has a lovely one in the queue, primed and ready to go: For many months he's been planning his first purchase when the IPO goes through. It's just about time to call up and reserve... dozens of frozen turkeys. One for each Atari executive, to be delivered just before they leave for Christmas holiday, each with its own greeting card expressing Rob's appreciation for their hand in creating his current success. He wouldn't want to overlook his benefactors at this special time of a very special year.

Unbeknownst to young Rob, the Atari execs are thinking of him as well.

In fact, the big wigs at Atari have prepared a gift for the Activision's and the Fox's and the Coleco's and the Mattel's of the world, and especially for Imagic and all the joyful Imagicians. Something very special indeed. After all, 'tis the season for sharing our true feelings with those we think of most. And sometimes we share what's in our wallets too. In fact, Atari has recently been looking very carefully at their wallet.

Truth be known, everyone in the video game industry is interested in getting a look at that wallet. But the time for sharing that information is still many weeks away, and it's only 48 hours until Imagic's public offering is set to unwrap the Christmas present that will answer many peoples' prayers, including young Rob's.

To honor this most auspicious occasion, Atari, in an unprecedented act of generosity, decides to release their financial results early rather than make everyone wait. You see, as the leaves began to fall in the autumn of '82, so did the prospects and projections of Atari's leadership, but nary a peep was uttered… until now. The entire industry gathers round to bask in the glow of the Atari numbers. And to their collective horror, the numbers are bad. Really bad. About half a *billion* dollars bad.

So bad, in fact, that all the financiers yell "Oh my!" and run away. Suddenly there are no nattily clad Santas to be seen anywhere. And with no one available to deliver the gifts, Imagic's IPO was cancelled less than 24 hours before its scheduled commencement. Rob could hear a chorus of frozen turkeys shouting unkind remarks, proclaiming the advent of a very sad Christmas before thawing into indifference. And it was a very sad Christmas indeed… for the industry in general, and for the Imagicians in particular.

That's the story of how Atari slighted Rob Fulop, and Rob got even with Atari, and then Atari got even with Rob (plus many others). And all it cost was the collapse of the entire industry. There are those who find Atari's actions shockingly unethical, but not young Rob. He summed it all up to me thusly:

"Why would I expect any different? It was fowl play from the start."

[NOTE to Anyone Still Reading: I may be paraphrasing somewhat.]

THE B.I.G. STORY

As I mentioned earlier, the story of the crash is a big story. It gets a lot of attention from a lot of different people. And when enough people examine any event of consequence, it invariably leads to a quest for answers. Exploration demands explanation, that's just how our brains are wired.

The video game crash of the 1980s is just such an event. In the decades since, lots of exploration has been done, resulting in many and varied explanations. Here's a few:

New technology paradigm. Investor interest lost. Fractioning and factioning internally. Previously unseen product lifecycles. Corporate culture clashes. Beginners attempting the expert trail. Short term industrial thinking. It was just a fad. Too busy to improve our methods. Loss of innovative impetus. New management models slow to be adopted. Profits dominating quality. Legal innocence and ineptitude. Lost commitment to integrity. The advent of a new labor market. Old eyes on new problems. Inability for intra-company communication. Intolerance of colleagues. Misreading market signals. Lack of awareness…

Clearly there are myriad possible (and quite plausible) explanations for the crash. But beneath all of these are a few fundamental causative factors, the pillars upon which the whole collapse was built. These are the underlying causes that make this a big story. In fact, it's the B.I.G. story behind the big story.

And by B.I.G., I mean: Bluster, Ignorance & Greed. Three key undercurrents of the dynamics which truly killed (or at least severely maimed) the video game industry in the early 1980s.

Let's take them one at a time…

Bluster (aka Hubris) – When I think of Bluster at Atari, two quotes come to mind:

> "Success has a thousand parents,
> but failure is an orphan."
>
> COUNT CALEAZZO CIANO, ITALIAN DIPLOMAT, [PARAPHRASED] 1942

> "Yea though I walk through the Valley of the shadow of Death, I shall fear no evil... because I am the meanest motherfucker in the Valley."
>
> BRUCE H. NORTON, FORCE RECON DIARY, 1969

In any highly successful environment, lots of people want to feel they are an important part of the reason for that success. The longer the success continues, the more this confidence in their contribution increases. Over time, they may develop a sense of invincibility, as if no one can approach them or compete with them. If left unchecked, this can go beyond feelings of pride or confidence and cross the line into symptoms. There came a point at Atari when many people from *all* parts of the company fell victim to BMOBS. If you don't know what BMOBS is, don't worry, I just made it up. But it's a very important acronym when it comes to how things fell apart so aggressively. BMOBS (pronounced Bee-Mobs) stands for Believe My Own Bullshit Syndrome.

People tend to go with their strengths, whether or not those strengths apply to the current situation. When the only tool I have is a hammer...

One thing is for sure, when I do things with the hope of achieving a certain outcome, and then I get that outcome, it reinforces my belief that I was a causative factor. Whether or not that's true, I'm still likely to believe it. Here's another sure thing, when you hire someone because of their past performance elsewhere and you pay them a lot of money, you're not sending them the message they don't know what they're doing. You're reinforcing their confidence in whatever they think is the right thing. They may have known exactly what they were doing in the last job, but that doesn't mean the same approach will work in the new environment. BMOBS can be very dangerous for people who think they already know what to do (as opposed to people who seek to figure out what to do). Atari became a magnet for people who thought they already knew what to do. Many people at Atari wore the badge of bluster with great pride.

Ignorance – It's important to be smart about ignorance. Please don't confuse it with stupidity. Ignorance is a lack of information. Stupidity is an inability to use information. Another way to think of it is: Learning can fix ignorant, but there's no cure for stupid. Ignorance and stupidity are very different characters, but they frequently fall off the same cliffs.

There were very few stupid people at Atari, but there was a lot of ignorance. After all, we were on the bleeding edge of hi-tech entertainment. Working at the State-of-the-Art is a funny thing. There was a clever game engineer at Atari who had the best definition for it. He'd say: "State-of-the-Art means when it's broken, nobody knows how to fix it." That is so true. We also discovered an interesting corollary to this: "When it works, nobody knows what to do with it."

Ignorance isn't a bad thing. Everybody's got some, and it's the place where every new endeavor starts. The magic questions are these: Can you recognize your ignorance before it hurts you? And if you do, will you start learning in order to heal it? Here's another question: Do you think bluster improves this picture or not?

Greed – This is something we couldn't get enough of at Atari. Of the B.I.G. three, this is the only one explicitly called out in the Seven Deadly's. Greed kept Atari so busy milking the cash cow, they forgot to find it a mate and launch the next generation. When we're too busy to improve our methods, we're just digging a deeper grave. The need to create a new game console was a hot button issue in engineering. Management never took it seriously, though. They preferred to just keep milking and milking the VCS cash cow as if it would never end. When the flow began to dry up, they didn't bother to call a bovine lactation consultant because they were too busy running around with all the milk buckets. They simply continued milking and pretended all was well.

Well... while they continued to milk, problems continued to mount and the whole situation began to sour. In short order, money shifted from the goal to the solution. Atari kept trying to buy their way out of problems rather than solve them, which led to more problems to solve. Ultimately, the people who couldn't get enough got too much and things changed in unplanned and unpleasant ways.

Bluster, Ignorance and Greed. Whenever you see the three of them together, you know BMOBS can't be far behind. When I'm full of shit, unaware that it's shit and can't get enough of that shit, naturally I'm going to start believing my own bullshit. Their combination contributed mightily to every aspect of the crash.

There are so many takes on the crash; There are financial takes, media takes, gaming takes and entertainment takes. Since you're reading this in a world thoroughly populated with video games of all kinds, you know the crash wasn't the end of the industry. Here's a take you may not have heard before:

The Evolutionary Take – The video game industry is an insect (think of all the people who get so *bugged* by video games). I was there through the end of the larval stage and into the pupal stage, where it cocooned for a few years. Some mistook it for dead, until it reemerged, exploding into adulthood as a radiant butterfly, fluttering to this day in the hearts of billions. Hearts that beat with the excitement and anticipation of every next video game release!

Isn't that a nice take? But regardless of metaphor, the industry did crash back then, and it crashed *hard*. People got hurt. People wanted answers; convenient ways to tuck it all away in their minds. Like Nolan says, "A simple answer that is clear and precise will always have more power in the world than a complex one that is true." For some people, ease and comfort are more important than accuracy.

Every tragedy needs a face, and E.T. was a very tempting face. It did ultimately become the face of the video game crash, but it was more of a symptom than a cause. The true causes are many and complex.

"First product lifecycle" is a good summary of the practical reasons for the crash. Similarly, "B.I.G. & BMOBS" summarize the personal reasons.

After all, every human experience is a mixture of the practical and the personal. The circumstances and the players, you can't separate them. And for the fledgling video game industry of the early 1980s, that mixture proved deadly.

In retrospect, I came to believe the single most significant transition in the path toward industrial destruction was Atari's change in leadership. This sparked a culture quake which rumbled around for years until it shook the underpinnings of the entire industry. When Nolan left and the reigns were handed carte blanche to Ray, it was a shift from Rock Stars to Towel Designers, and that didn't work out well at all. I'll get back to the main story now but fear not, we'll take more detours down the road and get back into the crash.

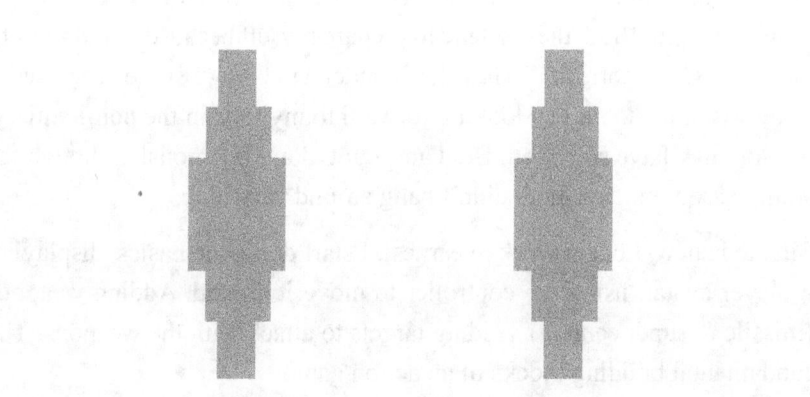

CHAPTER 5
ARE YOU GAME?

WORKING ON MY FIRST GAME

Shortly after receiving my first video game assignment, I find out I'm moving to another office. A desk has opened up elsewhere, and three of us in "The Sea of Desks" is apparently one too many. I bid farewell to Rob & Tod and move to the other side of the quadrangle. There I will share an office with a kindred spirit named Brad Stewart. When people hear I'm moving in with Brad, they tell me to prepare myself because "Brad likes to make puns. It's horrible!" They don't understand why I smile. They see it as imprisonment, but I'm looking forward to my time in the pun-itentiary. As you may have surmised, Brad and I got along infamously. Though for some reason, most visitors didn't hang around very long.

Situated anew, I begin work in earnest. I start doing the basics, displaying a player avatar, using the controller to move it around. Adding weapons (missile & super cannon). Adding targets to attack with the weapons. The fundamental building blocks of an action game.

The game that won't be Star Castle has no name at this point. In my head, I'm calling it "Time Freeze," connoting a spectacular payoff visual I'm committed to. It's an elaborate animation sequence, the reward for doing "the big thing" in the game. I'm not clear what that big thing is yet, but I know you do it with the super cannon and something amazing happens when you do. It looks absolutely spectacular… in my head. You may think I'm putting the sizzle before the steak, but you are absolutely right. It's the entertainer in me, which right now is entertaining the idea of a sizzling steak.

At its root, a video game only has two components: Graphics and Gameplay. Developers spend virtually all their time answering two basic questions: What will I put on the screen? (graphics) and What will I do with it? (gameplay). In the end, both must be satisfied, but where you start varies from project to project. I decided to concentrate on the graphics first. Glitter before substance.

With each thing I do, I'd ask myself: Is this cool enough? For instance, putting the player graphic on the screen. Some people start with just a rectangle and focus on the gameplay. They worry about the graphics later. For weeks or months of the early development, it just sits there while you iron out the game mechanics. That wasn't OK for me. I created an animation up front, so for all the hours I'm going to spend staring at this while thinking of other things, I'll at least have a more active display. It's not better or worse, it's just different priorities. I can't tolerate an inactive screen.

Every development has its ups and downs. You get ideas and try to implement them. Sometimes you can, sometimes you can't. If you can get them on the screen, sometimes they're fun and sometimes they suck. Ups and downs, round and round. Hopefully converging on a fun and engaging product.

At the outset things are going well, my glitter first approach seems to be paying off. I'm getting more dynamic visuals on the screen, which is drawing attention. More people are stopping by to find out what's up with my game. "That looks cool. What does it do?" My plan is working, except for one thing, I have no answer for them.

I've reached the point where there's glitz in the graphics but no pizzazz in the play. It's time to shift focus to gameplay. I take the basic elements of Star Castle, remove some, add others, then totally reorganized the screen geometry. One thing I keep completely is the controller scheme. Everything is coming along fine. I feel I'm making reasonable progress. But it's one thing to create all the game pieces, and it's quite another to play it.

There comes a point in every development where all the key rules and components are reasonably represented. They may not look good, but they work correctly. Now you can finally experience the actual gameplay. This is an important milestone in every video game. It is called (unsurprisingly enough) First Playable.

After a couple of months, I complete the most primitive version of first playable. Finally, I can sample my game. This is exciting, I pick up the controller and I can hardly believe it. I'm playing my game. It's all working exactly as planned. I play on for a while, then I ask a few other programmers to play it. Here is what I learn:

The play is cumbersome, it doesn't flow. Some interesting mechanics seem less accessible than they should. The controls are uncomfortable. I don't want to accept it, but my first game is lame. I need this game to be great… and it isn't.

Clearly, I'm going to have to tune it, whatever that means. I'm deeply disappointed. This is not how I pictured the progress of my original proposal to Dennis. But it gets worse.

After tuning it up a bit and playing more, it appears a tweak here and there won't do the trick. Something major needs to change and I have no idea what that is, which is even more disturbing. It's bad enough that it's lame, but what's really sinking my boat is the fact that this is my big idea. If I'd stuck to the original assignment, at least it could be the game's fault. But this is all on me because I *had* to make it *soooo* special. A victim of my own avarice.

I guess I need to relax my initial vision, solicit more collegial feedback and make some fundamental changes. But I need a break first, to clear my head and reset my thinking. Then some serious brainstorming with my colleagues should help. Fortunately, the perfect opportunity is coming right up. There's a companywide brainstorming offsite next week in Monterey. My first business trip! I have no idea what to expect, but I'm totally psyched.

BRAINSTORMING

At times when my head feels a little cloudy, I find myself hoping it blossoms into a full-blown storm. Brainstorms are the cornerstone of good design. If you accept the hypothesis that design is about ideas, then you must also accept that those ideas have to come from some place, and that place is a *brainstorm*.

Sometimes brainstorms are incidental, like aha moments in the shower or flashes of insight while I'm stuck in traffic. These are wonderful moments, full of excitement and relief. But they are hard to plan or count on.

Other times, we attempt to transform "brainstorm" from a noun into a verb. We put on our best thinking clothes, grab our idea bags and go brainstorming. At least, that's the goal. When we plan to arrive at new ideas, our reservations are not guaranteed. But brainstorming sure seems to increase the chances, right? Lots of managers agree, especially non-creative managers in charge of creative product developments. "Hey people, let's get together and be creative. You spout the great ideas and I'll write them down. Ready, GO!"

At Atari, we had a variety of brainstorming activities. There were the informal spontaneous types which took place around the office, in a bar or at a party. Then there were the formally sponsored and scheduled brainstorms, where large groups would meet off-site, somewhere nice for multiple days.

Two questions come to mind: 1) What is it like to have someone pay to send you to a beachside resort, cover all your expenses and the only requirement is coming up with some ideas which might help create products to entertain people with new technology in fun ways that will generate profit? And 2) What is it like to read a run-on question?

These formal brainstorms were typically held in giant conference rooms. The seating was arranged in a huge circle to make it more "interactive" and ideas were supposed to be presented. These weren't quite as productive as the small spontaneous get togethers. Not that ideas weren't generated, but the atmosphere wasn't terribly conducive. You are expected to tell everyone your idea while people who have little understanding of our product (and no vocabulary for the technology) sit in judgement. This would go on for a few hours each day, and the rest of the time was allocated to small group break-out sessions (code for goof-off time). I could never tell if these off-sites were seriously aimed at finding new ideas or if it was more of a boondoggle type fringe benefit of the job, because they did not seem to produce a bounty of useful product ideas. They did, however, generate some amazing moments.

My first official brainstorm came when I was only two months old (in Atari years). It was held at a beautiful seaside resort in Monterey, California. Most of my elder colleagues were used to this set up and less eager to contribute, but I was bubbling with enthusiasm and itching to make a mark, so I popped with an idea or two. I'm not sure they were well suited to either 1982 technology or the tender sensibilities of upper management, but what the hell? I tried.

One idea I proposed was a thing I call the Video Vignette. I liked the idea of being able to set personality traits in various characters and then turn them loose to interact with each other and see how it plays out. Though not a breakthrough concept today, in 1982, with the graphic capabilities of the VCS system, it was most likely a bridge too far. I liked the idea of creating scenarios and watching the drama, but if anyone would have said, "great idea, run with it," I wouldn't have had the slightest notion how to proceed.

My next idea, which was proposed half tongue-in-cheek and half tongue-elsewhere, was based on the directive that we should try to incorporate other aspects of the Atari universe in our product development plans. Here's a little-known fact: In the early '80s, Atari had the most advanced holography lab on earth. Holograms were the coming killer app, and although we could make 'em we had no idea what to do with 'em. Well, almost no idea. My proposal was to make full size human holograms on mylar, a material that might easily adhere to glass or tile. If we can get popular women and men to pose (preferably naked), then we can market "Shower with the Stars" holograms. Or at least get unknown men and women (preferably attractive and naked) and we can replace the word "Stars".

They rejected this idea as too unsavory, even if we used a front company to shield Atari or Warner from seepage. I think they were forgetting one of the fundamental rules of media: Pornography is frequently among the first profitable applications in any new medium. At least my concept fell squarely under the heading of "interactive entertainment," which was what they were seeking. After all, I am first and foremost a team player, at least by day.

[NOTE to the Aspiring Creative Manager: There are many books about brainstorming. Search for "brainstorming" on YouTube and you'll see titles ranging from "Brainstorming Done Right!" to "Brainstorming is bullshit!" (actually, this latter one is excellent). I've researched and practiced generating ideas my entire career. Allow me to save you some time with this summary: When it comes to brainstorming, people will share many techniques about how to map ideas, collate ideas, avoid discounting or criticizing ideas and coax brilliance from the masses, but these are just techniques. I believe the real answer lies in the raw materials. If you want to improve your brainstorming sessions, make sure you have at least one or two authentically creative people. Then harvest the ideas however you like, secure in the knowledge you've planted the proper seeds.]

Now, let's turn our attention to the after-hours activities which occur on off-sites such as this one, because during this brainstorming in Monterey, there was one unforgettable night that I will (by definition) never forget. It was another eye-opening introduction to this remarkable world in which I am now firmly entrenched.

One thing I came to understand at Atari, there's no way to anticipate the next drama, the next delight or the next left field moment. You simply know it's coming and hope you'll be there when it happens. In the world of Atari moments, Tod Frye is a carrier.

TOD FRYE CREATING BALLBLAZER.

PHOTO BY: DAVE STAUGAS

Though only weeks old in my new job, I have already sorted out my colleagues to some degree. Feeling good about my choice of who to follow, we set out for the promise of a solid evening's entertainment. Tonight will not disappoint.

After a long day of brainstorming, concept sharing and examining competitive product, we gather for dinner. As is often the case at Atari, this begins with drinks. Eight of us hit the bar a time or three, then seat ourselves around one of many large circular tables in the dining area. We're having a good time and chatting enjoyably when all of a sudden Tod bangs his now empty glass loudly on the table, startling everyone. As all eyes turn to him, he screams:

"T…I…S…"

The convivial din of the dining room ceases instantly. Concerned patrons in every direction turn toward our table. Their eyes anticipating some manner of explanation (oft the case with Tod). Those at our table aren't quite so shocked since it is not unusual to hear Tod announce his latest acronym. For instance, we all know MOO, Master Of the Obvious, which Tod proclaims himself to be. Then there's VOB, a Tod classic which stands for Victims Of Biology, which he believes us all to be. VOB was reserved already for the title of Tod's yet-to-be-created software company (VOBSOFT) or it was to be the name of Tod's going-to-happen-soon band. He even had a song written already. The lyrics go something like this:

> *Brain damaged baby, but that's OK,*
> *I had my E.S.T. today.*
> *Electro-Shock Therapy that's what I mean,*
> *Electro-Shock Therapy it's really keen.*
> *Electro-Shock Therapy it's good for you,*
> *Turns your brains to mushy goo… etc.*

Anyone who has spent more than a little time engaging with Tod is familiar with this song.

The point is: We are not as shocked as others in the room by this outburst, albeit his intensity level tonight surpasses his usual. We are, however, no less curious since "TIS" is a new one on us. Fortunately, the wait is short as Tod continues at full volume,

"T…I…S……Tod…Is…Drunk!"

After about three seconds of dead silence, our whole table bursts out laughing uncontrollably and now the evening is officially ON! I'm reminded of Bette Davis in the film classic "All About Eve":

"Fasten your seatbelts. It's going to be a bumpy night."

We laugh and drink our way through dinner and now it's time to retire to one of the hotel rooms for the kind of refreshment better consumed in a less public venue. The crew at this point consists of roughly seven of us, including mostly programmers and one budding department director with an ever-watchful eye on becoming VP. Let's call him "Jack." Jack is a pretty fun guy when he lets himself go, and tonight no one is doing anything likely to increase inhibitions.

We're on the fifth floor of a very nice hotel, having a lovely time enjoying the view through the glass doors opening to the balcony.

Having partaken some already, Tod is out on the balcony doing his acrobatic horsing around routine as he is wont to do. We are all in here trying to keep the smoke inside for discretion's sake, when Tod decides to take it up a notch (which is essentially what Tod is all about). He jumps up on the railing of the balcony and starts doing a tightrope routine. At this point one must appreciate that he is pretty high (in every sense of the word) and is taking quite a gamble. We are all equally elevated and enjoying the show, with some remote part of our diminished consciousness contemplating some of the poorer possibilities just inches away. Tod is showboating and loving it while we are laughing along. Somehow this all feels perfectly normal in context, which is the magic of Atari. Finally, Tod looks down from the balcony and says, "You know, this doesn't look so bad…" and he jumps.

That's right. He actually jumps off the fifth-floor balcony. In the down direction. For the second time this evening we are stunned and rendered speechless by an act of Tod. But this one has a substantially larger component of "HOLY SHIT!"

Faces change. Eyes speak of fright, worry and some empathy for poor Tod. But Jack's eyes tell a different story entirely. He is picturing the inquest at which he must explain how, as the ranking partier present, he thought it wise to play along this far with the troops (what can I say, Jack's eyes are very articulate). As the catatonia wears off, we rush to the balcony railing and look down, fearing the worst.

But all we see is Tod looking right back up at us, laughing hysterically at our frightened and agonized expressions. It turns out there's a plaza level (with some lovely shrubberies) built out of the back of the hotel on the fourth-floor level. A soothing wave of relief cleanses all as we realize Tod has only jumped a few yards. But he sold it like five floors, and we all bought it, elevating the drama intensely, another Tod specialty. One to which I aspire myself on occasion.

But we're not done yet. Tod begins meandering around the plaza, chortling beneath our flushed and slightly less horrified expressions. Suddenly he decides it's time for a bathroom break. So, he wanders over toward one of the bushes and prepares to relieve himself, right there on the plaza level topiary.

Jack, who hasn't said a word since Tod's original departure (and in point of fact, has only just resumed breathing), puts on his "management" voice and issues one of the classic quotes of Atari secret history. He points at the bush and yells, "Tod, STOP! You can't pee in those bushes. We might have customers in there."

Two days later, I ride back from this auspicious offsite with Jack. We are both ambitious people, and we like to plan big things with the absolute intention of making them happen. Several interesting ideas come out of this ride which lead to future projects. Interesting that the ride back generates more product ideas than the brainstorm itself, but that's how it is with the muse. Unresponsive to orders, then suddenly when you're not looking… BAM!

Jack is an interesting and accomplished person. He would soon provide me a key insight, forcing me out of my comfort zone and into a very successful game project.

AFTER THE CHANGE

Back at the office, I'm still stuck with a game that isn't working. What did I learn at the brainstorm? I guess I can jump off a balcony, but that's not really the solution I'm seeking.

The main problem seems to be the controller scheme, one of very few elements that remained true to the original coin-op game. Star Castle used the standard 4-button Asteroids style set-up (rotate right, rotate left, thrust and fire). The Atari joystick has 4-postions (forward, back, right, left) and one fire button. So, the firing is easy, that's the button. Then I'll just use stick-right and stick-left for rotation and stick-forward for thrust. That leaves stick-back free, which is handy for loading the super cannon and all my game mechanics are covered.

But it feels cumbersome and doesn't translate well to my new screen geography. Jack (who would soon become my manager, since bosses can change anytime at Atari) suggests a direct-motion setup, where the player just moves in the joystick direction. Sure, it could feel nicer, but it presents two problems: 1) How to control speed? and 2) How to load up the super cannon? If you can't load the cannon, you can't succeed in the game, which is a big problem. I cannot dump the cannon, so I dismiss Jack's suggestion. But it's not completely gone. Sometimes people tell me things I'm not quite ready to hear. I keep those words in the back of my head until I'm ready to take them out and revisit them. Jack's idea was exactly that.

But it doesn't take very long for me to pull his idea out and reconsider things. This is facilitated by the fact that the game sucks as is. Interestingly, this poor controller scheme will lead me to a few conceptual breakthroughs in game design.

Since development is a trial and error proposition anyway, I figure I can at least test the new direct-motion setup. If I don't like it, I can always hit undo and resume being stuck where I am. So, I do it.

It turns out the speed problem is no big deal, I just set a flat speed when you move and stop on release. It's very responsive, and the player motion is more accurate. I also get to dump all the physics calculations needed for variable speed. Inertia be damned! Reality is unnecessary, a game only needs to be fun. It's always good when we improve a game by taking things out instead of putting things in.

But the larger issue remains: Now that I'm out of buttons, I can't load the cannon from the controller. How do I do it?

After a bit of rumination, it hits me: Why not *play* your way to getting the cannon? The screen contains the monster (main target) sitting on the extreme right side behind a shield composed of many bricks. Originally, you shoot away the shield to expose the monster, then you kill it with a super cannon blast. What if, instead of shooting the shield, you bump up against the shield to remove a brick? And what if doing this loads the cannon? That could be cool. But what if you run out of shield? How can you load the cannon then? Well, how about touching the monster? This meshes beautifully with another design tradeoff I'd made. In order to attack the player, the monster occasionally morphs into its own super weapon, sparing me the need for another graphic element (which I don't have). Now, in order to load the cannon for the big payoff, I must bet my life on the timing. This creates a lovely and well-motivated balance within the play: If you want bigger returns you have to take bigger risks. Art imitates life!

Making these two changes brings the game to life. Suddenly, people are really excited about it. The transformation is amazing… as is the enormity of my relief. Now I have a solid base in both glitter and gameplay. I've got something worthwhile in hand and can simply concentrate on making it better. I'm over the hump of "Can I make a game?" What a tremendous boost to my confidence.

This is a pivotal moment in development, both for my game and myself. Things are looking decidedly brighter now. But here's a bit of Atari wisdom: Whenever you find yourself basking in the golden light of sunset, it behooves you to remember the darkness will follow shortly. Unbeknownst to me, another major change is occurring around me. One that has huge implications for how I see the world beyond my dev station.

Just as I'm turning the corner and my game is coming alive, a big thing happens: Dennis the manager, as well as Bob Smith and Rob Fulop (two of my dearest Atari friends) announce they are leaving. They are forming a new company called "Imagic," which will be another competitor, like Activision. This hits me on several levels.

I'm excited for my friends Bob and Rob, who are setting out on a new adventure. I'm also sad to see them go. I haven't been here very long, but we've shared a lot of good time together and I will miss them. At the same time, I can't help feeling like I'm being left behind, and right when things are getting so good with my game. I know the environment will be diminished by their absence. Wait, did I just say, "it's not as cool as it used to be"? OMG, it's happening already!

But that's Atari. The ups and downs walk hand in hand, and the hits just keep on coming.

This also resets my thinking about my interactions with Dennis. I wonder: When did he know he was going? I always thought it was very cool of Dennis to let me blow off Star Castle and do what I wanted to do with this game. In fact, it was very cool of him to hire me in the first place. But six months later, when I see he's leaving the company, it occurs to me maybe he didn't really care that much because he was on his way out anyway. When he didn't want to hire me, I talked him into it. When he wanted me to do Star Castle, I talked him out of it. Was that an acknowledgment of my passion or apathy from a short timer? It's the old retroactive recalculating of perspective, which I found a tad vexing.

Sometimes I think I'm so clever, when in actuality I'm just misreading the tea leaves. One thing's for sure, I never try to change anyone's mind about leaving Atari. I take solace in the knowledge I can go back to my office and trade puns with Brad Stewart. That's always fun. A few weeks later, Brad announces he too is leaving to join Imagic. Sadly, this will Stewart in a new era of people being warned about the *pun*ishment of sharing an office with me.

Meanwhile, back at the game, I refocus my attention on the glitter. There is a huge empty space in the center of the screen, just begging for something interesting and colorful. It's as if my garden is doing fine, but there's one spot that's absolutely screaming for a water feature. My inner economist has an epiphany, conceiving of an elegant way to splash some color down the center of the screen for visual balance. I say "economist" because this effect is an inexpensive way to generate a dynamic effect. How inexpensive? It cost me absolutely no additional memory space in the game. The epiphany is realizing I can use the computer code *as* the graphics.

[NOTE to the Non-Nerd: VCS video game cartridges contain computer memory chips that hold two things: Graphics and Computer Code. It's like a model airplane kit. The Graphics are the unassembled model pieces, the Computer Code is the paper assembly instructions, and the cartridge is the box. (True Fact: the kind of code we use to program the VCS is called Assembly Code) Just like a person uses the instructions to assemble the model from the pieces, the computer uses the code to assemble the game from the graphics. Now, when producing a model airplane kit, the pieces and directions combined must be able to fit in the box. Once the box is full, you can't add any more pieces. The same is true for video games. The cartridge (box) won't get bigger, so anytime I want to add something to the game (graphics or code), I must first make sure there is room available or I can't add it.]

Here's the cool thing about this approach: I get the graphics for free! Adding a visual usually means creating the graphics (bits which represent the shape and color of an object) and then storing those bits, using up precious memory space in the game. But this way I don't need *any* extra memory because instead of creating new graphics, I'm using computer code that's already there. Think of it in terms of the model airplane metaphor: If I'm good at origami, I can simply make the airplane out of the instruction sheet! Then I can throw away the pieces, still have an airplane, *and* get extra room in the box for adding new stuff to enhance the product. That's the elegance of using the computer code for graphics. This is what programming the VCS is all about, a never-ending quest to overstuff the box.

[NOTE to the Nerd: If this whole model airplane metaphor isn't making any sense to you, it's probably because you skipped the last Non-Nerd note. If you go back and read that, this will likely become more meaningful for you. But, as in life, there are no guarantees.]

And there it is, I've gained an interesting visual that requires no additional memory for graphics. It's a win-win! It will, however, go on to create a legal hassle for management. It's always something.

But this technique would prove more than just a colorful water feature and copyright headache. Throughout this project, I've maintained the goal of creating some spectacular effect. It's been relegated to the back of my mind because the front has always been occupied with survival. Now that things are moving in a more positive direction, my thoughts increasingly return to my grand payoff vision. After some trial and error, I happen upon a variation of the code-for-graphics technique which allows me to spread this effect over the entire screen. Add a few animation steps and some synchronized sound blasts, and voila! I have the first full screen explosion in the history of home video games, which is nice.

Now people are loving the game. This is huge for me, because it's proving out my initial design. Like many designs, my basic concept was solid but raw. It needed a few magic ingredients to realize the potential. Fortunately, I had good people around me to help hone that initial idea into a playable game. It… is… ALIVE!

This process taught me some valuable lessons in video game design:

Listen earnestly to feedback. This is not as easy as it sounds. Sometimes suggestions can feel like a threat to my vision or my implementation skills. Beware ego-impaired hearing, and overcommitment to concept. A plan is merely a basis for change. It helps to remember it's not my job to create everything, but rather to make the best choices from all the possibilities. Of course, this advice applies well beyond video game design.

Also, anytime you take a game mechanic off the controller and put it in the play, the game gets deeper and better. In some games, button mashing IS the gameplay, but generally speaking, it can be more fun and rewarding to use a play sequence instead of a button. This advice is likely of little use outside of video games.

I continue adding enhancements and ramping up difficulty as the game progresses. Finding atypical ways of approaching other aspects of the game is how I'm spending most of my time now.

Things are looking good for my first game. So much so, that the winds of release are beginning to blow. The time has come to turn my attention to that which I have given the least thought: Naming the game and defining the setting (or theme). I need to give this some attention now because other people are starting to give these topics some attention. Those people are from…

[ominous music riff plays]

… Marketing.

WHAT'S IN A NAME?

For most of the development of my first game, I've put the sizzle before the steak. At my root, I want to dazzle players and give them something amazing. It's the entertainer in me. It's the marketeer in me. The conflict between engineering and marketing grew to account for a significant part of the Atari work experience, but it also plays out inside my brain at times. And it came to a head when it was time to finalize my first game.

We've reached the point where I have a good game, but it still has no name, which is fine. In my view, you don't start with a label. You start with interesting tweaks and techniques to maximize the hardware impression. Then you make sense of it later with a game "setting" and possibly a storyline. That's my approach to Yars and all my games for that matter. Of course, the issue of what to call the game would not come into play on my next two products.

What became known as a "Yar" was my first graphic, and initially it had nothing to do with a fly. It was simply something with moving arms that might look cool and symmetric when animated. My original working title was "Time Freeze," but that disappeared along the way, leaving nothing but abstract gameplay and graphics in its wake. The concepts of Ion Zone, Qotile and all the Yars' Revenge lore would be a total retro-fit on already existing elements.

As in life, I do what I think is cool and then make up explanations for it later. That's how I believe we are wired. We act on our emotions and then we rationalize things to provide the illusion of making sense. That's certainly how I approach game design. My thinking was all about maximizing the gameplay and visual impression. Does it really matter what you call it?

Things look good with the game and it's time to start thinking about shipping it. Naturally, marketing needs a game title to put on packaging and promotional materials. The no-name status cannot continue. Marketing typically takes care of this. They have their own creative process for generating a name and packaging, as well as assembling all the product details.

But the thing is: I want/need to make everything about this game the best it can possibly be. I can't just leave the naming to marketing at this point for one very simple reason: I'm a control freak with tremendous confidence in my creative capacity. It's hard to release creative control to other people (especially people in whom I lack creative confidence). So, to answer my own question: Yes, it matters what you call it. It matters to me.

I ask the marketing rep if I can make my own name submission. He says, "Fine, but do it soon." I tell him to come back tomorrow morning and I'll have something for him. I spend the rest of that day and all night in my office trying to come up with the best name I can.

I wanted something simple but sharp. Compelling and intriguing without being off-putting. It should connote action in a way that's inviting to a player. Hmmm. It occurred to me that "Revenge" is a great title word because it tells a whole story by itself. And who doesn't want revenge? OK, I'll grant you there may be monks in Tibetan ashrams who have no vindictive feelings whatsoever, but they are not the target demographic so I'm going with it. The next obvious question is: Who is seeking revenge? I must name my avenger. This, I realize, poses a very interesting prospect…

One thing I've always wanted to do is add a word to the English language. It's one of my many dreams, and this is my golden opportunity. Everybody knows what Pac-Man is. If my game is a huge hit, the character name I choose right now could become common parlance. Have you ever tried to make up a word? I start reeling off possibilities, but everything sounds stupid. This is my big chance and nothing is working. I'm frustrated.

Eventually I give up on sounding good and start thinking about alternative ways of approaching the problem. This calls for an algorithm! I need a systematic way of choosing a name. It occurs to me it should be more than a name, and my mind starts spinning up. The name becomes a concept, then a theme, and finally the whole thing congeals into a marketing plan. My marketing plan to market my plan to marketing. And here's how my plan goes…

I'll use a cipher. The name will be an encoding of something else. Something irrefutable. Something like… Ray Kassar, the CEO of Atari. Yes! The title of the game should be Yars' Revenge, with Yar being Ray spelled backwards. What's Kassar backwards? Rassak? OK, the game will be set in the Razak solar system. But that's not enough. What makes a name stronger for marketing? A package. How can I make Yars' Revenge a package? I know, I'll write a story to go with the game.

So, I start writing, and by the crack of dawn I'm staring at 12 handwritten pages entitled, "The Yarrian Revenge of Razak IV." I'm too tired to realize it, but I've just created the first backstory in video game history. It's an action-packed jaunt through space, detailing how Yars came to be and how they came to inhabit the Razak solar system. This includes how they were wronged on such a grand scale that they are now motivated to dedicate their entire race to seeking revenge. I finish tweaking it about 7:30am-ish, just in time to hand it off to the freshly arriving admin for typing and I go catch a short nap on the floor of my office. Around 10am the marketing rep shows up and I give him my proposal for Yars' Revenge. The game's afoot.

Later that afternoon, the self-same rep comes strolling by and informs me my proposal is officially submitted and under consideration. Time to activate phase two. I thank him and offer to share an insider insight if he'd care to hear it. This piques his interest. I insist this is top secret and must be held in strictest confidence, lest it influence the outcome unduly. He assures me he shan't tell a soul.

"OK then," I say, "You know the Yar in the title?"

"Yeah."

"Spell that backwards."

He thinks for a moment and says, "Ray?"

"Right. And how about Razak?"

After another moment or two, "Kazar? Ray KASSAR?!?!?! Does Ray know about this?"

"Of course, I wouldn't do this without his knowledge. But you can't tell anyone. I don't want it to influence the decision."

I swear him to secrecy twice more and send him on his merry way. At this point I know three things: 1) He's going to run straight back to marketing and tell everyone. 2) No one in marketing has the stones to broach this with Ray Kassar, which is a good thing, because 3) Ray knows absolutely nothing about this. I've never talked to Ray Kassar. In fact, at this point in my Atari life, I haven't even been in the same room with him. It's a pure bluff. I'm feeling pretty good about this as I head home for some needed rest.

The next day, in walks the marketing rep, "We're going with Yars' Revenge. Congratulations!" His face beaming with delight.

And at that, I start beaming a bit myself.

Now my game has a name, it is officially Yars' Revenge forever more. I'm feeling a deep satisfaction, believing I've outmaneuvered marketing with my clever little ploy. This satisfaction would last many years, until I realized something which changed the picture substantially. Another of my retroactive recalculations.

I was so proud of having pulled off my master plan, I never considered the possibility my concept might have been good enough on its own, which is kind of sad. I was so busy trying to pull one over on marketing, I didn't stop to think maybe I was only pulling one over on myself. But here's the thing about fooling yourself: It's really hard to tell when you're doing it, because no matter how smart you are, your brain knows all the best hiding places.

While I'm standing here with the marketing rep, Ray still doesn't know anything about my maneuver, but he'll hear about it soon enough. Of course, I won't find out he found out for another five months, when Ray and I end up chatting briefly at a press demo for the upcoming Christmas season. It'll probably go something like:

> Ray emerged from the slew of media people crawling around the room. He approached me and said, "Hello Howard, I heard about what you did with Yars."
>
> "Yeah? What did you think about that, Ray?"
>
> He half-smiled, "Just keep making games, Howard."

You already know this because we visited that future a long time ago. Perhaps you did a retroactive recalculation of your own. But for my own part, I'm just right here right now, so I don't know this yet since it won't happen for five months. Still, I'm pretty confident about my prediction. That's the beauty of non-linear timelines.

But something has to fill the five months between now and when I talk to Ray at that press demo...

How about answering the unanswered question: What legal issues were raised by my glitzy center-screen water feature?

Using computer code *as* the graphics is an interesting idea, but it also creates an interesting problem: Putting graphics on the screen is no big deal. Anyone can look at the TV and see how a Space Invader or a Pac-Man is drawn, but putting the code on the screen is another matter entirely.

Think of a Kahuna Burger. Graphics are like the bun and meat patty, there's not much mystery about how those are created. The code, however, is like the secret sauce in the video game recipe. You don't want to reveal those ingredients. Putting the code on the screen sounds like you might be giving away trade secrets, and that is a dangerous thing to do in business.

Naturally, when the lawyers hear I'm displaying the code on-screen they freak out. I don't blame them. They've been aggressively pursuing Activision and Imagic in court, trying to retroactively protect their intellectual property. Then here comes Howard, laying it out for all to see. This leads to hours of spirited discussion, whiteboard diagrams and a brief presentation with hand puppets. Ultimately, they are satisfied I've guarded the secrets adequately and they sign off on my colorfully controversial display. Another thing I learn during this exchange: My job is way more fun than theirs.

As you may have surmised by now, I'm no stranger to controversy. For some reason it follows me around. That's not a total coincidence, of course. Though generating controversy is not my intention (well, mostly not), it does seem to be a frequent result of my style and approach to life in general.

This is likely because I rarely ask: "How do I do this?" I'm more focused on: "What am I trying to accomplish?" Then I start imagining possibilities, and my imagination tends toward the path less traveled. Whatever the task, I'm likely to do it differently.

One aspect of the propensity to do things differently: People don't always relish the consequences. When something is truly new, it pushes us beyond our usual repertoire. This can irritate some people who end up dealing with it. I'm familiar with this phenomenon.

For example, the legal issue raised by displaying the Yars' Revenge code on screen. No one had done that before, so it never came up. When it did, the lawyers had to deal with it. They did not share my enthusiasm for this innovation. But it all worked out in the end.

Another example is the hot button issue of credit for game authorship. Atari steadfastly refused to give it. Under no circumstances would Atari reveal the identity of any internal game developer. To the outside world, we were nameless Atari employees, period. Activision and Imagic made it a selling point, putting the programmer/designer profiles right out front. This fact was not lost on Atari engineers... but some Atari engineers were lost to that fact (when they quit Atari and joined those competitors). Yars' Revenge broke that streak. I wasn't the first to seek credit, but I was the first to receive it. Not by asking for it, but by creating one of my unexpected solutions.

No one had ever seen a back story for a video game. But when they did, the marketing gears started to turn. They were inspired by my backstory and decided to turn it into a comic book which would be "packed out" with the game. This is called an ancillary product. It was the first of its kind in video games.

Well, Atari had published comic books and magazines before, just never as an ancillary product. That turned out to be crucial, because this way the magazine rules took precedence over the video game rules. What does that mean?

One of the fundamental rules of magazine/comic publication is: Always include credits for the people producing the product. Not so with video games at Atari. Since they had to do credits for the comic book, and the comic book was ancillary to the video game, they needed to include a credit for "Cartridge Programmer," followed immediately by a hyphen and then my name, complete and spelled correctly. This was the first time any Atari programmer got credit for their game in the product. Mind you, it wasn't on the box or in the manual, it was in the ancillary comic book. But it was there, and I was pleased to take it and did not complain.

OK, I had one complaint. In the first proof of the comic, I noticed someone else had the credit for "Story." I asked why this was so, since I had written the story. The product manager said: "Look Howard, you can have one credit in the product. Do you want the story credit or the programmer credit?" I found this to be a very reasonable response to my query and took the programmer credit. In the final version of the comic, they dumped the "Story" credit and replaced it with "Writer," which was a great choice. Hope Shafer significantly adapted the original story for the comic book. She absolutely earned and deserved a unique credit. For the record, illustrators Frank Cirocco, Ray Garst and Hiro Kimura deserve major kudos, as does art director Steve Hendricks (whose name was misspelled "Hendericks" in those same credits). Fans tell me how much they love the Yars' Revenge box art and illustrations. I've always been proud to be associated with these images.

For my first game, Atari asked me to do a coin-op conversion. What I heard was: "Please make us a salable product." I ended up giving them an original game as well as the basis for ancillary products and a new licensing property. I made a contribution that wouldn't be denied, forcing the issue of credits in a new direction.

I enjoy giving people what they're asking for, but not what they expected. Sometimes it's more than they expected, which frequently delights them and makes everyone happy. Other times I like to play Monkey's Paw by giving people precisely what they asked for but NOT what they wanted. You know, like computers do on occasion. I find people are less enthusiastic about this version. It's a fun game for me, but not so much for the other players.

This isn't just an Atari thing with me either. In college, though I graduated with two full majors and an additional minor in just three years, the Dean still gave me shit because I qualified for summa cum laude but refused to write a thesis.

[NOTE to the Non-Academic: When you graduate from college, you get a diploma. But wait, that's not all you get. A Bachelor's degree is only the base camp. As you climb GPA (Grade Point Average) mountain, there are several plateaus of note along the way. If you do well enough, you can get "cum laude" added on. This is Latin for "with honor" and it's a lovely little lagniappe you get automatically for good grades. For those who climb a little higher there is "magna cum laude" which means "with great honor." To earn this distinction, you must write a thesis, which is a major research paper. And for the extra-motivated climber who approaches (or reaches) the summit itself, there is "summa cum laude" which means "with highest honor." To get this, you must write the same thesis the magna's do *and* you have to "defend" it. This means convincing a panel of professors you didn't just cut & paste it, but actually understand what you're writing about in your thesis.]

I happily settled for "cum laude" because it came with no additional surcharge. Writing a thesis and defending it? That one extra word on my diploma wasn't worth the hassle. "But you're cheating yourself if you don't dive deeply into your subject matter," insisted the Dean. I countered, "But what about breadth? And the value of multiple perspectives?" He insisted that deep is the only way to go, so I went... without. I took what

I'd already earned and left the rest on the table. I felt sufficiently honored already and saw no need to gorge myself to excess at the academic honors buffet (to be fair, I do hear their word-salad is so good it defies description).

Though I maneuvered well through college, I really wasn't an academic at heart, much to the chagrin of the Dean. He presumed my membership based on my performance, but ultimately rejected me based on my approach.

Has it been five months yet? No? OK, let's revisit another question: Does it matter what you call a video game? More specifically: Does setting/theming influence the overall play experience?

After all, gameplay is gameplay. Can graphics and theming alone significantly modify a game's impression? Let's see…

Take a simple pong-like gameplay. Here's the design: You control a paddle which moves side-to-side near the bottom of the screen. Objects fall from the top of the screen and you must touch them with the paddle before they reach the bottom. The player is rewarded for touches and penalized for misses. That's the entire game design. Issues of speed, timing, number of lives, etc. are relegated to tuning. There is no theme or backstory. You can play this game without any additional information or graphic enhancement. Your motivation is to increase your skill level for this abstract task.

Now let's try applying two different themes to this gameplay. Remember, all we are changing is the graphics and the backstory...

Theme 1: Someone is throwing babies off the roof of a burning building and you must save them by catching them on the pillow you're holding (the paddle), thus returning them to their families. Same game, different motivation. Now you are a hero!

Theme 2: Someone is throwing babies off the roof of a building you just set on fire. The families are waiting below, hoping to catch their falling babies. You are just above the families, catching the babies in your gaping mouth (the paddle), eating and digesting them, then pooping on the families below (bonus points for this). Same game, different motivation. Now you are a monster!

This one design yields three entirely different play experiences (and probably different ratings), yet the gameplay remains exactly the same.

Is there one you would purchase? Is there one you would protest? Does setting/theming make a difference?

Are we there yet? Wow, waiting five months can really test your patience. It's testing mine, largely because the main thing filling these five months is testing. Yars' Revenge consumer testing...

RELEASUS INTERRUPTUS

Have you ever been involved with consumer product testing? It's a very interesting experience and quite a learning opportunity. It certainly will be for me.

Yars' Revenge, my first game, is about to undergo more consumer testing than any other game in Atari history. I'd like to say there are many good reasons for this, but in truth there is only one: Someone keeps demanding more testing. My first game is building a real fanbase as it nears completion, but it has its critics too.

My excitement builds as the game release approaches, but someone raises a concern about the game's appeal and a Focus Group is ordered. This is where eight players come in, play the game for about 30 minutes, then eat pizza and discuss the game while interested parties (myself included) watch from behind the mirrored glass. Yars does fine in the focus test. OK, now we can release the game.

Things start moving forward again. Suddenly there is another concern about the game's quality. Another test is ordered. Yars does well and seems like a solid candidate for release. In my opinion, Yars is done. In fact, I'm already working on my next game, Raiders of the Lost Ark, but I really want to see my first game released. I'm itching to taste that Atari magic.

Once again, everything looks good. Yars seems to be in the pipeline, when someone starts claiming there are "long term playability" issues. At this point I'm developing a bad case of Releasus Interruptus. Yet another focus group is commissioned. Yars passes with flying colors. OK, third time's a charm. We should be clear for take-off now, but just as we start revving the engines, the control tower waives us off yet again.

Apparently, there are still some "concerns" about the game. Really? Now I'm wondering what the hell is going on?! It appears someone has an issue with Yars' Revenge and they have the ear of upper management. Don't these results mean anything?!?! After all this testing, I'm getting a little testy myself.

Internally, my game is getting significant recognition and some real love. Developers & Marketeers are admiring and playing it. There is even some interest in making a coin-op version, which has never happened before with a VCS game. Oh baby! I'm poised on the verge of release, ready to join the ranks of proven game designers. To be a member of the club. But I'm not quite there yet. I've always been happier as a free agent, the outsider, even in situations where I had a choice. Atari is the first time in my life being an insider feels important, or worse yet, necessary.

Releasing Yars' Revenge will make it all happen. It's my graduation, coronation and rite-of-passage; The Acceptance Trifecta. I'm in the rocket, ready for blast-off, but the countdown keeps getting interrupted. Not long ago I was too depressed because nothing was working. Now it's working beautifully and I'm too anxious.

As you can tell, my case of releasus interruptus is becoming more acute, though I assure you there is nothing cute about it. This really hurts. So, what do you take for that?

They decide to go hard core and commission a Play Test. This is the Rolls Royce of video game testing. Whereas a focus group takes about two hours on some afternoon or evening, a Play Test requires an entire weekend. Over 100 players come from all over the demographic spectrum. Each one will play and evaluate two games, the target game and a control game. Naturally, Yars' Revenge will be wearing the target, but a Play Test is all about the control game. I'm dying to know what I'm up against. I hope it's not *too* good.

Finally, word comes. The control game will be… Missile Command. Oh, SHIT! Missile Command is a tour de force. It's the best game to date on the VCS. And it was created by none other than my good friend Rob Fulop (you may remember him from Christmi past). And that's not all, the test is being held in Seattle. They're going out of town for the big showdown?!?! I immediately book my trip, because there is NO WAY I'm going to miss this.

Apparently, all I have to do is outscore the top game on the system and Yars' Revenge will be released. WTF? Does it really have to be this hard? They say if you want to be the best, you have to beat the best. I believe this is true. But in all honesty, I'd be okay winning against a second-rate title and just getting out the door. I'm an optimist at heart, but I'm not feeling too confident as my plane touches down at SeaTac airport on Friday evening.

After a tense night, Saturday morning arrives and away we go. I'm securely ensconced in our little booth behind the mirrored glass. It's like a bunker. I'll be here for nine hours today and nine more tomorrow. This is my weekend, sitting in a hot sweaty box in Seattle with several others, watching people play the two games and waiting for the scoring sheets to come in. I happen to have a bachelor's degree in statistics, so I am poised to start tabulating those sheets and working the numbers, baby. The first score sheet arrives. They love Missile Command. They do not like Yars' Revenge. At. All. This does not bode well for my big weekend.

Sheet after sheet comes in. I'm counting it up and checking the numbers as we go. It turns out that first sheet was the worst one for Yars in the entire test. When the smoke clears (thirty-two hours later) Yars' Revenge beats Missile Command in the play test. In fact, it got the highest rating ever seen in a play test! This made the flight home significantly more enjoyable than the flight out.

Well… it turns out there are no more objections. After squeezing as much delay as possible into my delayed gratification, the game is cleared for release, this time for real. After seven months of development and five months of testing, I finally receive a dose of elixir. Yars' Revenge leaves the building and finds its way into well over a million homes. It goes on to become one of the most beloved games on the VCS. Oh. Baby.

My releasus interruptus drops into full remission. Releasing a game is the apex drug at Atari, and it's a powerful rush. Seeing commercials for your game on TV. Seeing it on the shelf in stores and knowing you created it. Reaching and entertaining millions of people. And the peak experience: Seeing kids fighting over the joystick for a turn to play your game. I get to feel all these things. What. A. Rush. This is the real drug at Atari, and I'm addicted. Now my mission in life is clear: Keep increasing my supply.

Yars' Revenge will always be my most cherished development. It established me as a credible game designer and programmer at Atari. Having this grand plan of what the game needed to be, and seeing those plans come to fruition. Feeling I've arrived and belong to a place I can call home. This is deeply moving for me.

I'm filled with gratitude. I also pick up some well needed experience points. So, what did I glean from this scene?

I learned that people can reliably tell you if they like something, but they cannot reliably tell you why. This suggests we are in touch with our experience but not so much with our internal process. It's the Black Box effect of the mind. What's a black box effect? I'll tell you a little later, when we pull back the curtain technology hides behind. That's coming right up in Chapter 7, Bob. I promise.

I also learned how secret agendas can come out of nowhere, causing real damage. Corporate politics can be brutal, but I think I'd rather engage in it than be subjected to it (a commitment which leads to many future lessons). Also, never underestimate the potential for a bad situation to limp along indefinitely.

Perhaps the biggest lesson of Yars' Revenge is the realization of value in myself. For the first time I can use my quirky skillset to create that special combination of providing entertainment while succeeding in a difficult task. This is supremely satisfying for me.

After the better part of a year at Atari, I've completed my first full game creation/release cycle (and I'm well into my second). I've contracted and cured a major case of releasus interruptus. I'm a member of the club now, and on my way to experiencing the ultimate Atari high. Finally, for the first time since I was hired, I can exhale.

Ahhhhhhhhhhhhhh.

JEROME DOMURAT & HSW CHILLIN' IN THE CAFE

PHOTO BY: DAVE STAUGAS

CHAPTER 6
WHAT'S NEW?

I'M READY FOR MY CLOSEUP

"Hold your breath and shut your eyes!" someone shouts, as another bold gust whips through, unleashing a mega-burst of micro-daggers, each one stinging on impact. The warning was muffled slightly by the bandana guarding the speaker's face, but appreciated nonetheless.

Having never experienced a sandstorm before, I giggled at first when the crew handed us our cliché dark blue bandanas. But clichés always come from somewhere. Now I understand why cowboys wear bandanas. They're not just props for old westerns, they're the hardhat of sandstorms. A piece of protective gear like safety goggles, which BTW also work in a sandstorm. You better believe I'm wearing mine, and much obliged for the option.

The powerful desert winds are spinning up dust devils, mini tornadoes of swirling mess. It calls to mind one of my favorite words: vortex. I'm here in the desert because I got sucked into a vortex, which in this case is not all bad. Though it's persistently pelting me with flying bits of garbage, this vortex is also connecting me with some remarkable people. Zak Penn, for instance.

Zak is a Hollywood screenwriter of note, having *Penn*-ed such notable films as Ready Player One, The Incredible Hulk and Last Action Hero, to name just a few. Zak is directing and starring in the film that brings us all here. He has previously directed two other entertaining films, and this effort is shaping up to be another.

Zak is the funnel cloud that scooped me up in California and deposited me here in the desert on this extraordinary April day in 2014. Zak is a formidable presence, tall and fit (compared to me), with a goatee, a congenial bearing and a sense of humor so dry it blends seamlessly into the arid New Mexican air. Interestingly, he's here in the desert because of a double Chinn. Simon & Jonathan Chinn, that is. They are the seasoned executive producers, who between them hold two Emmy's and two Oscar's for other documentaries. It's hard to say if this will be a major film, but there is major film talent on this project. It's exciting to be here with them.

Zak didn't sneak up on me though. About seven months ago I began receiving bulletins from the VGCEWS (Video Game Community Early Warning System). Each announcing the funnel cloud and wondering if I'd been contacted. "Not yet, but I'm sure I'll hear eventually." After months of this, I started having the authentic Hollywood experience of wondering: "Will they ever call?"

Zak and company reached out in February. They're doing a documentary about an urban legend. The legend alleges that millions of my E.T. video game cartridges are buried somewhere in the New Mexico desert. The project's working title is "Atari: Game Over." I like this, it has a nice ping to it. We had a lengthy phone conversation which led to several follow up chats which led to a sizable film crew (and *all* their equipment) jamming into my little dink-ominium for an all-day interview session which led to this run on sentence detailing the whole sequence. I gave them six hours of intimate details about my past and present. I really opened up, even sharing some thoughts and feelings I hadn't previously discovered. It was a very emotional experience for me. They did a great job.

It's an unsettling thing, handing all that footage over to someone and trusting they will do "the right thing" with it. I ask others to do this in my productions, knowing I'll do everything I can to maintain integrity with my interviewees. But when I'm the interviewee, I can't know the director's true intention until I see the product. Of course, past performance is usually the best predictor of future performance.

You may recall Zak Penn has directed two previous movies, so I rewatch them. The good news is: They are very good movies. Wonderful contributions to one of my favorite film genres. The bad news is: That genre is Mockumentary. Mockumentaries are fictional documentaries. They are usually satirical in nature, if not outright attack pieces. If Zak is looking to do a hatchet job here, I've basically spent six hours sharpening his axe. Fortunately, there's something about his questions that gives me hope this film is aiming higher.

Either way, here I am: A dufus in a dump in the desert. A Pollyanna with a bandana, surrounded by blasting sands and a sea of fans. How awesome is this? Very. All in all, my kind of vortex.

I'm here in this mixing-bowl-maelstrom because of a movie. But the movie is here in order to fill a vacuum created by an even bigger, more powerful vortex: the one created by Microsoft Corporation.

PIONEERING A NEW MEDIUM

Streaming media is the hot new medium and everyone is vying for market share. Microsoft is spinning up to enter the fray too, by creating attractive content for the launch of their new streaming video service on Xbox gaming consoles. And who will create this attractive content? Xbox Entertainment Studios. And who are they? To quote the Google search results:

Xbox Entertainment Studios was an American

television and movie studio based in Santa Monica, California created internally by Microsoft Studios in 2012, in order to create "interactive television content" for Xbox Live.

This movie, Atari: Game Over, is the first of six planned shows. Xbox Entertainment Studios is doing a series about the origins of video gaming and interactive media. Two points of note here: First, making interactive media about interactive media is delightfully self-referential. Second, isn't it interesting that they're using "was" in describing Xbox Entertainment Studios?

Regardless of what this dig may ultimately reveal, we've uncovered the original motive for what we're all doing here in the desert. It's all about exploring and establishing a foothold in this brave new medium of streaming content. And that's what we were doing at Atari, pioneering a new medium.

Back then, it was the interactive entertainment revolution. The new medium of Home Video Games was turning passive television watching into *active* video game playing, creating hot new interactive experiences on the very couch where you used to nod off. Change TV and you change the world!

Developing for the Atari 2600 was a very interesting exercise, because the goal wasn't just to do something; the goal was to do something no one has ever seen or done before. You had to think in different ways, you had to innovate. When that's the assignment, how do you start?

It takes a while to comprehend the potential of any new system. Consequently, the first thing people do in a new medium is replicate the best stuff from prior media. Naturally, the first thing we did was copy coin-op games, but this strategy couldn't last. Trying to keep up with advancing coin-op technology on the aging VCS is like trying to keep up with a 10-speed when you're riding a tricycle. It just isn't going to happen. That's where we came in. It was our job to figure out what was unique about this system and give it a life of its own.

I saw my job at Atari as more than producing video games, it was my job to make a contribution to video gaming. I was very aware this was a new medium, the first I'd seen in quite a while. Being a pioneer in a new field was tremendously exciting. At long last I'd found my calling, the thing that spoke to me. If I'm doing something fresh and new, then I feel good about my work… and myself.

Yars' Revenge broke new ground in a number of ways. You already know it's the first video game with a backstory, an ancillary product (the comic book), a full screen explosion and the first Atari game to credit the programmer. It had elaborate death sequences and an extremely intricate soundscape. It was not the first game with an Easter Egg (a hidden message in the game), but it was the first openly acknowledged, Marketing-approved Easter Egg. In fact, there was more collaboration with other parts of the company than was typical of games at the time.

Did you know that Yars was also the first video game with a pause mode? Pause is something you would never put in a coin-op arcade game because: Revenue! But no one thought to put it in a home game before this. The next time you take a leisurely bathroom break without losing any lives, say "Thank you, Howard."

My first game was a game of firsts. That was my aim. I fit right in at Atari because I wanted to be an innovator, needed to be. There were several reasons for this, but perhaps the simplest is this: Doing things the way they're usually done is boring… and I hate boredom.

That's another thing we were doing at Atari, alleviating boredom. Isn't that what entertainment's all about? To take arms against the vast wasteland of teen angst and ennui, that others might endure it more easily than had I. Now there's an agenda I can get behind. Although I admit, there are times when I'm thankful I didn't grow up with video games. I might never have left the TV.

We were working at the State-of-the-Art, blazing trails into unexplored territory. It's exciting, but it's also frustrating, even treacherous at times. There's a price for admission to State-of-the-Art.

When you're attempting breakthrough work, you can't know if your ideas are doable until you do them. Frequently you'll try things that don't work.

This presents a question: Is it impossible to do or am I not clever enough? Not skilled enough? It can mess with your head. Managing this anxiety is a crucial part of the creative process.

There are many egos in the entertainment industry, mostly because egoless approaches rarely produce great entertainment. To create something worthwhile I have to bring my whole self to it. My commitment to vision is a powerful source of energy that fuels my efforts. The people who really succeed at Atari put everything into what they do. Total commitment is essential. This may be a productive way to go but it is not a particularly healthy one. There were divorces and family issues. People became socially isolated, except at work. They developed personal problems and/or financial difficulties. The work becomes so consuming, nearly every other aspect of your life fades away. When I'm that invested, failures can be devastating.

When I'm truly bought into succeeding in a creative production, fear of failure can provide the energy it takes to push through some dark moments and arrive at remarkable results. That feels amazing. Sometimes a setback ends up revealing a better alternative, the breakdown leads to a breakthrough. But when I'm really pushing the limits, I hit a lot of dead ends. It's the thrill of climbing a dangerous cliff. Reaching the top is incredibly rewarding, but when you fall the pain can be enormous.

These games generated phenomenal amounts of money, which in turn generated tremendous pressure to produce and deliver them. The industry was so young, so dynamic. It was a hotbed of potential and a forge of stress. We all felt the heat. The light that burns brightest burns shortest, and eventually burns out. Atari was the Mount Olympus of burnout. Some bent under these conditions, others broke.

There were more nervous breakdowns at Atari than anywhere else I've ever worked. One programmer disappeared for days. Upon visiting his apartment, they found the door wide open. Inside, the poor guy was sitting on the floor of an empty room, unresponsive. Another programmer went catatonic in his office, just staring at the screen. For some, the stress was too much. Some people got carried away and others got carted away.

Atari was a challenging workplace psychologically. One day could be rainbows and the next could threaten pink slips. It could flip in a moment, and regularly did. It was not a place for the faint of psyche.

But for a budding psychotherapist, Atari was an amazing internship. Nearly every kind of neurosis and disorder was flagrantly on display. Sometimes insanity is your greatest business asset. This is balanced by Tod's observation about our lifestyle at Atari. As he put it, "It wasn't really much of a foundation for a healthy human life."

That's the thing about the entertainment business, it attracts a certain personality type. It's not just the ego, though that's there. Validation is a major part of it. The sweet fruit of validation pushes people beyond their limits. Happily, we volunteer to jump into the fire.

It's a risky business if you take it seriously. Why do it? Why put yourself through this? Perhaps the best answer comes from my long-time friend and Atari colleague Rob Zdybel. He was the one who taught me the fundamental trade-off of video game production. Rob explained to me how making a video game on the Atari VCS system typically demands more than a thousand hours of obsessive strenuous labor; hard thinking, stress & tension, blind alleys and lost sleep. This is not fun. It's draining and it takes a toll on your life. But if the game does well and reaches hundreds of thousands of people, and each of them plays the game for just ten hours, then your thousand hours of work ends up generating millions of hours of joy. That's a pretty good return on your investment. Amen to that, Rob.

The process of making entertainment, the nuts and bolts of bringing it all together, is a lot of hard work. But every once in a while, it leads you to a place of pure magic. That's the allure of the entertainment industry. That's why we do it.

And in video games, I don't have to be beautiful. It doesn't matter if I can sing, dance or act. If I can program, I have a shot.

A typical day at Atari was so atypical. Our job description: Do something cool no one has ever seen before. A successful day at Atari ends with something that didn't exist when the day began.

Now I have to deliver a game five times faster than anyone ever has, and

I still want to make it a good game. It's clear 1,000 hours is not going to happen, but I will jam well over 500 hours of work into these 5 weeks.

CHAPTER 7
A NERD WORLD COUNTRY

"PRETENSE BUILDS WALLS. GENIUS OPENS DOORS."

Deserts may be the last unspoiled horizons on earth. Mainly because they are inhospitable to human life, so why hang out there? But people usually find a way to use what they have, and one great use for desert land is garbage burial. That's another reason we're all here.

I can't help noticing the towering yellow giants cranking away, their motors roaring as they work. Huge limbs groaning and grinding as they move massive amounts of sand and garbage out of the way, dipping ever deeper into this dark tunnel to the past.

Imagine if a small crew of people had to accomplish this without machines. How long would that take? How many times would they have to tie someone to heavy ropes and gently ease them down into the hole? How long could someone spend down there filling pails with dirt and curiosities? Waiting for the pail to be hauled up, emptied and returned. How many pails would it take to fill one of those big yellow claw-buckets with the silver teeth? The big yellow monster just grabbed and dumped another thigh-high pile in less than 60 seconds. This gas-fueled giant is converting dead dinosaur juice into awesome digging power. That's what technology does.

THE COUPLE THAT DIGS TOGETHER.

Technology is the common name for the product of homo sapiens' proclivity for toolmaking. Technology makes the impossible possible, and the possible more efficient. In Stanley Kubrick's 2001: A Space Odyssey, there is a scene in which some of the early release versions of human beings first discover they can use a bone as a tool (and ultimately as a weapon, though that's really just another tool). This moment from the dawn of toolmaking suddenly transitions to a space station docking sequence, brilliantly conveying more than ten thousand years of technological evolution in a single cut. It's one of the most powerful moments in film history. Likely because it's one of the most powerful statements about human history.

Here in the desert today, we are witnessing the use of one technology to help reveal the history of another. I'm writing this book to share a bunch of stories and ideas with you, many of which revolve around technology and the hi-tech corporate workplace. Why? Because Atari was a major inflection point in the technological history of our society. Essentially, we're visiting a nerd world country, where you'll be hanging out with me and my silicon friends. To get the most out of your travel experience, it's going to be important to understand a few things about nerd world culture and issues. There are lots of ways to think about technology and technologists.

It's also true that in the world of technology it's easy to talk over people's heads. There are nerds who delight in locking others out of the conversation with technobabble. They take obscure terms and nomenclature, mix them up in a big bowl of mumbo jumbo gumbo and spill it all over your nice clean clothes. Then they steal your napkin and laugh at your dry-cleaning bills. This is very discourteous and only serves to separate them from you. Some nerds are like this. I hate to judge, but these are bad nerds. More accurately, they give the term NERD a bad rap. Shame on those shameless naughty nerds.

I prefer to build bridges. I understand not everyone is a technologist (or even a nerd for that matter), so I want to offer a simple and clear explanation of what many consider a gnarly topic.

I'm going to tech-splain some basics about tech and techies, and along the way I'm hoping to bring a bit of humanity to the nerdly point of view. After all, nerds are people too.

So, let's share some fresh insights and perspectives about the world of technology and its inhabitants, thereby increasing the likelihood that, as the book continues, we will remain on the same page.

[NOTE to the Non-Nerd: Relax. You do not need to know anything about technology or programming to understand what I'm going to explain here… at least that's my ambition.]

THE BLACK BOX EFFECT

Technology is an interesting thing. Technological innovations usually make some part of our lives easier or better, but they do so by making the solution more mysterious than it used to be. Take person to person communication for example…

Talking is a fine way to communicate with people. But once they get beyond yelling range, I need another way to communicate with them. Enter technology: The pencil and paper.

Now I can write a message and send it to my friend. This makes communication possible, and I totally get how letter-sending works, but it's a very slow conversation. Enter technology: The telephone.

Now I can have a full-on conversation with my remote friend, which is great. And I understand how to do it, too. I simply speak into this part and listen at that one, but I have no idea how the words go from here to there. I no longer understand how this works. But here's the big question: Do I care?

For most people, the answer is NO. As long as I can use it, I don't need to know how it works. I don't have to understand the physics of a combustion (or electric) engine in order to drive a car. Imagine if you needed to be a civil engineer specializing in sewage systems in order to use a toilet. Remember the cartridges we plug into the VCS to play a video game?

They were just plastic boxes holding computer memory chips, but players used to call them tapes. Who cares? No matter what you call it, the game plays the same.

When a system's technology is hidden from users without reducing usability, the place that tech is hiding is known as a Black Box. It can be an actual box like a stereo tuner casing, or a conceptual box like the postal system. Either way, black boxes let me get what I want without having to understand the functional guts. This is a huge advantage... as long as it works.

Here's the downside: If it stops working, I'm stuck. Do I open up the black box and try to fix it myself? Or must I find an expert who can do it for me? Those experts are always happy to help, but they don't want to do it for free.

Every black box poses a mystery. How complex do I reckon this tech to be? How long do I think it would take to learn it? Do I believe it's possible for me to learn it? Do I even want to?

The answers to these questions constitute the Black Box Effect. I know the black box is hiding the technology, but I don't know how complicated that tech is. The Black Box Effect is a measure of the perceived complexity, it's an expression of just how complicated I *assume* the innards to be.

The word "assume" is very important here. Because until I study the underlying tech, I don't know how challenging the problem is. All I have is my estimate of the complexity which may or may not be accurate. But that's fine, because it's not about accuracy.

The black box effect is about impact. The greater the black box effect, the more helpless I feel when dealing with the tech. Therefore, the more I'm willing to pay someone to do it for me. The black box effect determines the pay scale of experts for any given black box.

My point is: Computers have a huge black box effect! Computer technology can be rather intimidating, leading most people to assume computers are significantly more complex than they actually are. And thank goodness for that, because the main byproduct of this misconception is dramatically higher pay for computer programmers, hardware engineers and IT consultants.

No one is immune to the black box effect. Before I began to study computers, I assumed they were more complicated just like most people do. Yes, I too once sat on that side of the wall of knowledge.

Once I started studying, I was both relieved and perplexed. On the one hand, I was very glad to find they weren't nearly as complicated as I feared. I mean, if anything was ever designed to just make sense, computers are it. All they have is logic and rigid rules. Predictability is their entire raison d'etre.

On the other hand, I was a bit shocked and discombobulated over the idea it was all so elemental. It was a kind of reverse black box effect. At times I felt I was dumb or losing it in some way because this can't be as straightforward as it appears. I must be missing something. This is more a statement about me than computers, of course. I was so sure they'd be hard to grasp that I doubted my actual experience. Expectation is a powerful force, but rarely a helpful one.

Most things are not as complicated as they initially seem when I view them from the right perspective. Despite that fact, I still felt intimidated by the prospect of learning computer science. If I stuck with this first impression, I'd keep it at arm's length and never come to appreciate how approachable they really are.

The hi-tech industry remains grateful to the majority who don't venture in, you've made our careers a far richer experience!

INSIDE THE BLACK BOX

Computers have become a major part of our world. A tool for business and pleasure, an icon of science fiction past and dating apps present. Yet most people don't really understand computers. The good news is: It's a lot simpler than you think.

A computer is just a machine, like a lawnmower, except instead of having a motor, wheels and blades, its power comes from a "processor" and its moving parts are 1's and 0's.

Computers seem bossy, but they don't mean to be. It's simply their job to collect info from some things (input) and then tell other things what to do (output). For instance, a computer might collect temperature reports from thermometers, and use that information to tell the central air system whether it should make things hotter, cooler or just leave it alone.

The computer takes input and processes it, then it outputs commands to other devices. This is why the computer's brain is called a "processor." But computers don't have a mind of their own, all they do is follow instructions.

That's where programmers come in. They are the computer's boss. Programmers create detailed instructions which tell the computer how to "process" the information (input/output). They communicate with the computer in its own dialect, known as the computer's Programming Language. Every programming language has a set of simple instructions that are clearly understood by the computer.

Computer engineers use these instructions to tell the computer how to perform specific tasks. They do this by putting sequences of these instructions together in a very specific order. When they're done, the sequence of instructions they created is called a Computer Program. Computers execute (or "run") programs. That's all they do. Sometimes when computers are broken, they also double as door jams or paper weights, but working computers just run programs.

When you play a game on your phone, all your doing is asking the computer in the phone to run the program associated with that game. When you send a text, you are running a different program. Apps, games, email, printing payroll checks, google searches, sexting, getting driving directions... these are all just programs running on the computer in your phone or tablet.

A computer program is like a house made of Legos, where each piece is one computer instruction. When all the pieces are together in the right place, you can see the house is done and it looks right, inside and out. Similarly, when all the instructions in a computer program are in the right place, the computer does exactly what it is supposed to do. No more and no less.

Although a computer is like a lawnmower, programming a computer is not like mowing the lawn because the act of programming a computer is not a physical act. Believe me when I tell you: It's totally mental. In fact, it's an exercise in communication. The programmer decides what they want the computer to do, then tries to translate this idea into the computer's language. If they do it right, it all works as expected. But funny things can happen when mistakes occur in the translation process. Sometimes the computer does something it was not supposed to do. Sometimes it does not do something it was supposed to do. When either of these happen, it's called a Bug.

In a Lego house, a "bug" occurs if pieces are missing or wrongly placed. Spotting a Lego bug is easy because you can look it over and tell if it's put together correctly or not. Bugs are easily fixed by simply adding, removing or rearranging Lego pieces.

Computer programs are like this too, except you can't just look at it and see if it's OK. You have to "run" the program to see if it does what you want. Computer bugs are far less obvious than Lego bugs. And if it's a complex program, some bugs can hide in very remote places where programmers don't always look. Some bugs aren't discovered for years because they're very good at hide and seek.

People like to blame the computer when bugs happen, but it's not the computer's fault. The blame lies with the programmer.

Oh? Why is that, Howard?

It's because computers are perfect, and people are not.

Really? But perfection is so rarely achieved.

Yes, that's true. It's a miracle of modern science to be sure. But the fact remains that computers do one simple thing, and that is: Follow Instructions Precisely. That's all they do. Programming languages are not designed to be easy for humans to use, but rather to be absolutely unambiguous for the computer to understand. Which leads us to the fundamental truth of all computer programming:

Computers do exactly what you tell them to do, which may not match what you intended them to do.

A bug is merely a difference between what we wanted the computer to do and what it actually did. Of course, the computer only does what we tell it to do. This forces programmers to confront the idea that we aren't always as clear as we think. Some programmers do not enjoy this confrontation. Nonetheless, a bug is a communication error committed exclusively by the programmer. Why?

Because computers won't read minds, nor will they make assumptions. They will not add meaning or interpretation. They will simply carry out the program instructions precisely, on that you can rely.

When humans converse, the phrase "you know what I meant" comes up occasionally. This is because we expect others to not only hear what is said but also to interpret meaning accurately. You know what happens when you tell a computer "you know what I meant"? Nothing happens, not even a bleep-bloop. Unless the computer is programmed to recognize this phrase, then it will do whatever the program it's running says to do when it detects this phrase. Every time you hear someone say "you know what I meant," you can rest assured the speaker would have created a bug if they were programming.

I hope it's clear at this point that programming computers is all about communication. Therefore, computer programmers must be skilled and attentive communicators to do their jobs successfully.

So, if programmers are such practiced communicators, why is it such a pain in the ass to talk with them? OK, I understand your frustration. After all, I've created it many times in the past.

It's important to understand that computers have no tolerance for ambiguity. Once programmers spend enough time socializing with computers (i.e. programming), they tend to develop the same intolerance.

If you're having trouble communicating with a programmer, here's what might be happening: You may be demanding that they make the proper interpretation of what you're saying (after all, you know you're being clear since you know what you mean), and they may be demanding that you say things precisely so there's "no room for ambiguity in the interpretation" (read: ability to blame me for doing what you asked). It may not sound like much but it's the basis for a hell of a conflict. Neither party wants to be responsible for the problem. What problem? The one which inevitably comes about when developing and marketing software products.

In nerd-world countries, it's all about learning the lingo. Nerds are linguistically quirky. They like playing with language more than the average person since that is how they make their livelihood. It can feel like they're creating linguistic barriers. In most cases they're not meant as barriers, it's more like an ever-evolving set of secret handshakes to the programmers' club. We love it when you join in.

PROGRAMMER BRAIN

A computer program running on a computer is a system. Programmers who create, modify and repair them are systems engineers.

Systems engineers analyze systems to determine their capabilities and limitations. Then they figure out how to best achieve goals within those systems. This is something all human beings do. We are all systems engineers, it's just that most of us don't do it explicitly with specialized language. For example:

Whereas a systems engineer might wake up and say, "Hmmm, I've discovered a bug in my set-up procedure last night, resulting in a current circumstance outside the range of acceptable outcomes. To recover to a nominal situation, I must derive and execute an optimal (or near optimal) strategy for transporting myself from my present location to a predesignated alternate location in minimal time. This strategy must assimilate and accommodate the current state of all relevant environmental parameters."

Most other people would say, "OMG, I forgot to set the alarm. I better haul ass to work. Oh $#*%, traffic!"

Same situation. Same reaction. Different language (and more of it).

But programmers are more than systems engineers; they are also a rapidly growing segment of the population. They show up more frequently in our day-to-day lives, so it's probably worthwhile getting to understand them a bit more.

Computer programming is a complex conceptual undertaking which happens entirely in the mind. Where else can engineers build working systems without building muscle mass? On the upside, worker's comp claims only come about from trying to move the keg. On the downside, the lion's share of my effort is invisible to others.

Programmers spend their lives trying to handle every possibility and doing so as efficiently as possible. Computer programs must handle ALL possible inputs, which means programmers spend their time trying to think up every perverse thing that could possibly happen, regardless of likelihood. This includes the absurd, the incongruous and even the ones they don't know about or can't imagine. Every miss is a bug that may reflect poorly on the perception of their skills. This is where their heads live. Ever try planning for unforeseen events? It can be very stressful. When your head lives in this place too long, it can lead to a phenomenon known as Programmer Brain.

When you only think in terms of exceptions, you can lose touch with ordinary happenings and the regular flow of life. Many programmers fall into this trap. They're so busy trying to protect the software from extraordinary cases, they lose their sense of proportion. Most people focus on the usual stuff, figuring they'll deal with the exceptions as they arise. This works pretty well, since common cases are the vast majority of life. It's not so good for programmers though, because they have to deal with the exceptions before they happen.

Good programmers tend to be people who like order and love exceptions. They love exceptions so much it can make them annoying. They love to point out the one obscure counterexample that your otherwise sound explanation doesn't cover. Programmers find this game amusing. Who doesn't love a good round of "Spot the Logical Flaw"? Many non-programmers, that's who.

Programmers may use more words to describe a situation than most people do (as illustrated previously in the wake-up example). This is due to their need to be specific and exact in their communication. Which leads to another important point: As previously noted, programmers aren't big fans of ambiguity. Ambiguity leaves room for mistakes and blame, neither of which appeal to programmers. Ambiguity also makes it harder to do their job, which is forming reliable strategies within systems. In fact, the easiest way to break down the effectiveness of most systems is to introduce more ambiguity (also known as chaos or entropy).

Programmers also have an interesting relationship with rules. They both respect and can't stand rules. Specifically, they like sensible rules and will not tolerate nonsensical rules. Of course, nonsensical rules are defined as rules they don't like. They don't phrase it that way, they say those rules are outdated, illogical, contrary to stated policy or just stupid. But once you translate the nerd-speak, that's the bottom line.

Because they have such an intimate relationship with rules, programmers tend to relate to systems differently since systems are defined by rules. Programmers may view solutions differently, and may enlist unusual approaches. Not necessarily to innovate but just for expediency at times. For example:

There was the time several of my programmer friends decided to go to a movie. They further decided it would significantly enhance their moviegoing experience to be drunk during the film, but there were no bars near the theater. They didn't want to risk sneaking alcohol into the theater, and they didn't want to drive while drunk. What they did was drink several alcoholic beverages very quickly, then jump in the car and drive to the theater as fast as legally possible. Their plan was to arrive at the theater before actually becoming intoxicated. They found a way to achieve their objective without breaking any rules of the system. They were able to enjoy the movie buzzed, never drove drunk and didn't sneak anything into the theater.

You may be thinking: "Big deal, that's not what I call a productive application of problem-solving skills." And you may be right, but it is an innovative approach that meets all the criteria and solves the problem. You may dispute the value of finding clever ways to be drunk in public places, but this is the same kind of thinking that leads to finding ways to make your car safer while reducing its cost (which, by the way, is another thing these same engineers do). Of course, they get paid to come up with innovations that provide real value to people, and no one was paying them to come up with the theater solution, but that's the point. Whether you pay them or not, this is what's going on inside their head. All the time. Programmer Brain is pervasive. There's no turning it off and there's no putting it away.

Programmers are also optimizers. In addition to handling every possible input, they seek the most elegant solution. Toward this end, software people are constantly spouting, critiquing and reforming ideas before settling on an approach. They're chasing the question: "Is this the best approach?" (Fun fact: The efficiency of a programming effort is typically measured by how nicely it solves the problem, not by how long it takes to create the solution. Planning time may be ignored entirely by a programmer, but not by their manager.)

Once a path is finally chosen, the second-guessing and reworking begins. Warning: Doing this for extended periods may have side effects. The impact ranges from the comic to the tragic. My wife sums it up nicely with: The Cabinet Story…

For her it was simple, "Sweetheart, could you please move the office cabinet to the den?"

"Absolutely." Done.

25 minutes later she comes by the office. I'm striking a pensive pose, staring intently at the open (but as yet unmoved) cabinet.

"What are you doing?" she asks.

"I'm thinking about the best way to do this."

"You take the books and games out, move the cabinet, then put them back in. That's the best way to do it."

"But where do I put them in the meantime? Should I move them to the den as I take them out or move them later? Do we really need all this stuff, maybe this is the time to dump some of it? Oh, there's backgammon, we used to play backgammon, that was a lot of fun. Or maybe I can carry it without having to remove everything. What's the minimum stuff I can take out and still move the cabinet? Maybe we should get a hand cart, that will save time on loading and unloading."

"Save time? Seriously!?!?! It's been half an hour. You could have done it already!"

"Look, planning time is just overhead. I'm trying to optimize here!"

She disappears down the hallway, but her sigh of exasperation lingers. In moments like this I'm glad she loves me, but this is what it looks like when Programmer Brain kicks in. It's a place where my approach to the task becomes more important than completing the task.

Remind you of anyone you know?

Programmer Brain isn't just an occasional quirk, it's a full-on world view and a fundamental way of being. It becomes the systems analyst's system of analysis… for better or for worse.

STARING INTO THE TECHNOLOGY MIRROR

Lots of people can teach us about computers, but computers can teach us about people, too.

Computers force us to confront the vast chasm between our intentions and our words. They accomplish this by doing exactly what we say, regardless of what we mean. They show us how much we expect listeners to make up the difference.

Another thing computers teach us is this: The biggest change in the world is the change from 0 to 1 (or from 1 to 0). That's certainly true for computers. The only requirement in a computer's life is being able to reliably tell the difference between 1's and 0's, everything else follows automatically. Every change in a computer's world is between a 0 and a 1. But our human world is not so black and white. We have lots of other numbers beside 0 and 1. However, if you ask parents with multiple kids which one made the biggest difference in their life, the vast majority will tell you it was the first. The reason the testing phase of Yars' Revenge was so hard on me is because I was looking to go from 0 to 1 in released games. "You never forget your second time," said no one, ever. Going from 0 to 1 is the biggest change.

We can learn even more about people from computer video games. If you see Bob, please tell him we'll get to that soon enough.

CHAPTER 8
GAMES & THEIR MAKERS

In the Venn diagram of life, the world of Video Games is a tiny circle that sits inside the overlap of the Technology circle and the Entertainment circle. In other words, video games are a small part of two giant worlds. So far, I've been speaking broadly about technology. Now let's refine the discussion by confining ourselves to the tiny circle…

WHAT IS A VIDEO GAME?

"Those who know, do. Those that understand, teach."

Many people think this is a misstatement of another quote:

"Those who can, do. Those who can't, teach."

This second quote, which many consider to be the actual quote, was written and published by George Bernard Shaw in 1905. It is definitely a real and recognized quote. But the first one isn't a misstatement, it's a real quote too. It was written by Aristotle, more than two thousand years earlier. Shaw was satirizing the wisdom of Aristotle, yet Shaw's quote is much better known. It's also more cynical. That's the thing about entertainment, it's catchier and spreads faster than long standing lessons. Sometimes comedy overshadows wisdom.

That's too bad, because wisdom has real value for us. Let's hang with Aristotle for a moment. I think he's saying that it's one thing to do something, but it's another thing entirely to understand what I'm doing. This becomes especially important if I want to do it better. Think about Black Boxes, Aristotle may be distinguishing between those who know what's inside the black box and those who don't.

We all know what video games are. But if E.T. showed up tomorrow and asked you to explain video games, what would you say? Think about it for a moment. There are lots of familiar things in our lives that we never have to explain because everyone already knows what they are. Explaining them to someone who doesn't know can be challenging, largely because we've never had to do it. Being aware of something and understanding it are two different things.

A video game is first and foremost a computer program, but it's also a game, a puzzle, an audio/video experience and a test. It's interactive and it's *supposed to* be fun. It's all these things and more. How would you explain this to someone who doesn't understand "fun"?

Fortunately, we know what fun is (or isn't, if you're more of a negative-space type). Right now, I think it would be fun to share some of the ways I think about video games. Let's see if they match yours...

There are three primary ways I think about video games. The first one that springs to mind is the software perspective. As a game maker, I see video games as computer programs; collections of computer commands laid out in a very specific order, just like millions of other programs. But a video game is a special kind of program. A video game is what's known as a Microprocessor-based Real-Time Control System. I know that might set off your jargon alarm but bear with me, it isn't as bad as it sounds. In fact, you probably know what this is already, you just didn't know there was a name for it.

Let's break down Microprocessor-based Real-Time Control System. "Microprocessor-based" just means the computer we're using is a tiny one that fits in small places. "Real-Time" means the system responds immediately when anything significant happens. There cannot be any delay between when a need is detected, and the required action is taken. "Control System" means there is some job or function that the program is performing. Think of a thermostat. If the temperature goes below 65 degrees, the heat goes on right then. The heat must keep running until the temperature is 65 degrees again, then it shuts off. If I hit the fire button in a video game, my avatar must shoot immediately. Back in 1981, there were two primary applications for this type of programming. Guidance/targeting systems for the military and video games. In my view, entertaining people was infinitely preferable to killing them for twelve cents a head.

When you program Real-Time systems, you learn some interesting things about time and people. In order to be Real-Time, the video game must respond instantly to any player input. But what does instantly mean?

From the player's point of view, instant response means no *perceptible* lag between hitting the button and seeing the action occur. Any delay at all feels uncomfortable and distracts from the play experience. In human beings, this typically means less than 1/30th of a second. Just to be safe, let's say the game must respond within 1/100th of a second. That's pretty fast, for a person.

Now let's check it out from the computer's point of view. In a modern video game, when the human manipulates the controller I have 1/100th of a second to respond. Cool. That means I have time to do 50,000 things, then after that I can do 300,000 more things, then I could do another *two million* things, and still have plenty of time before I need to respond to that wacky sapien. In fact, I could throw in another million and the player still wouldn't notice any delay.

Computers are incredibly, unbelievably fast (it just doesn't seem like it when you're waiting for one). They operate in a time frame beyond our comprehension. In grad school we wrote a simple program to do a few thousand things and then light up a display at the end. When asked how fast the computer operates, the professor told us the program will be done running before we finish pressing the start button, which was true.

[JOKE for the Nerd: Did you hear about the new Cray computer? It's so fast it can run an infinite loop in two minutes!]

[JOKE for the Non-Nerd: Did you hear about the nerd who thought that was funny?]

This doesn't mean humans are slow. We are simply low-resolution instruments. That's why we build tools to assist us. We use microscopes and telescopes to see better. We use microphones and speakers to hear better. Computers are tools we use to hasten our information handling. What cars do for our legs, computers do for our data.

The computer on the Atari system is more than a thousand times slower than the one in your phone, but it can do about 25,000 things before responding "instantly" to the player. It's a different scale of experienced time. Working on Real-Time systems can really mess with your sense of proportion and urgency.

My second take on video games comes from my inner phenomenologist. In addition to being a Real-Time Control System, a video game is also an interactive environment that monitors you and responds to you. At its root, a video game is a biofeedback loop.

Here's how it goes: Your fingers manipulate the buttons and levers on the game controller. The controller sends signals to the console where the processor interprets the signals and, according to the content of those signals, updates the video display and audio output. Your eyes and ears detect the updated audio/video and send these updates to your brain (*your* processor). Your brain interprets this new information and sends out fresh signals which flow through nerves down your neck through your arms into your fingers which then update the controller and the circle is complete. It's a biofeedback loop. You're feeding the computer by responding to what it's doing, and the computer is feeding you by responding to what you're doing. Around and around and around it goes... until you need a food break. Perhaps something with chicken and eggs?

A video game is an environment that responds and reacts to you consistently, which is one step beyond a biofeedback loop. Now we're entering the world of environmental simulation. It's fun to immerse myself in realized fantasy worlds and play in them. But the same tech and expertise used to create a fantasy role play game can also be used to train people for real world jobs, or run experiments with reliable results at a small fraction of the cost. Real-time environmental simulations hold incredible value for humanity. This is one of many ways video games (and their progeny) are changing the world.

My third perspective on video games is a nod to the performer in me. A video game is an entertainment experience. This accounts for my adherence to the principles of movie making in game design and creation. A video game does not take place on the screen, it takes place in the mind of the player. It's less about the game and more about the game *impression*. This is the basis of most decisions I make while creating a video game. I'm not programming to the machine. I'm influencing the player's senses and perceptions.

A video game is an entertainment application of a real-time control system. Rather than trying to make a fun game, my goal is creating an intense experience in the player's mind. If I do it well, gamers will enjoy the ride and that should create a sensation of fun.

The phrase Broadcast Medium was also pinned to my brain. Every game is a "show" that gets broadcast to a potential audience of millions. What is my message for them? My game requires players to act in certain ways in order to succeed. What is that communicating?

The fourth of the three ways I view video games is as a psychotherapist. Video games are like personality tests, they can tell you a lot about their designers and players.

To understand an artist, look at their paintings. To understand an author, read their stories. If you want to understand a video game maker, check out their games. This will tell you plenty. Why? Two reasons: We are all unique and everybody signs their work.

When doing any task, I'll put something uniquely "me" into it. This will differentiate it from anyone else's version of the same task. The markers are there, you have only to read them clearly. Video game developers tend to make games consistent with their desires, goals and worldviews. In this way, they sign their work. Players tend to play games that reflect the life experiences they seek and the things they most value. I might go so far as to say if you really like someone's games, you may well like the person too.

As a therapist, I find these aspects and delineations quite compelling. I enjoy strolling through a gallery looking at the paintings and figuring out what the artist was like. What kind of person would make these choices for the canvas? I do the same thing with video games. I don't just enjoy playing the games, I enjoy deciphering what games say about the makers and the players who engage them.

There are many kinds of games; driving games, shooters, adventures, flying, platformers, targeting, treasure hunt, etc. There are also many kinds of kinds of games. I think of these as game styles.

Game styles are things like: Does this game have a finish or am I always playing toward a higher score? Is it a pattern game like Pac-Man or read & react like Robotron? Is luck a significant part of your success as in backgammon, or is it more deterministic like chess? Is it an action game, an adventure or a hybrid? Most games have some aspect of competition, but which one? And let's not forget the ever-popular anal-retentive games like Space Invaders and Asteroids. (Hmmm, should "anal retentive" be hyphenated?) What do these various styles say about their players and makers?

Do you prefer quest/mission-based games, or would you rather play the how-high-is-up style where your goal is simply a higher score? Some people like to see something through to an unambiguous finish whereas others like the ongoing challenge of doing better than last time no matter how well last time went. Put another way: Do you prefer the clarity of closure or the thrill of unlimited potential? The arrival or the journey? Both are fine goals, but they speak to decidedly different personal preferences. It's not about better or worse, they are simply aesthetic choices we make.

How about competition? Most games engender competition of some sort or another. Do you like to compete against other people, yourself, or an algorithmically based computer opponent? Do you like to compete as part of a team or individually? Do you like one opponent or multiple opponents? These are also different choices which reflect personal styles. Some are more socially interactive; some are more individually focused. Some have you anticipating your opponent's psychology (social challenge) and some have you trying to figure out the environment's algorithm (technical challenge). Are you a lone wolf? Are you a team player? Or maybe you are team captain! Who you are shapes what you enjoy.

At Atari we used to talk about anal-retentive games. Asteroids (not Assteroids) is the all-time classic of the anal-retentive genre. The player is confronted with a mess which they must clean up. Upon successful completion of this janitorial charge, the player is rewarded with a new (usually bigger) mess. Asteroids is a poetic example because in the process of cleaning it up the player must first make a bigger mess (splitting each asteroid into many smaller rocks) before dispatching them to oblivion. And since there is nothing else on the screen, a genuine clean slate is achieved. Some people love the satisfying nature of anal-retentive games. There is something about cleaning up a mess and seeing your progress so clearly displayed. It's great validation for gamers. I play, therefore I am! Just plug in your game Descartes and away you go. Enjoying this does not define you as anal-retentive. It might reveal obsessive-compulsive leanings, but it's more likely the simple joy of immediate gratification.

How about pattern games vs read & react? Pac-Man is a game where maxing it out means knowing all the patterns and executing flawlessly for screen after screen after screen. Robotron throws you into semi-randomly generated chaos and expects you to fight your way out every time. Are you Sir Lawrence Olivier, taking pride in delivering a perfect reading of a script? Or are you James Bond, shooting from the hip and dealing with whatever comes your way instantly and responsively? Let me answer this question with more questions: Do you like to live by a plan, or are you all about spontaneity? Do you like knowing where things are going, or do you prefer to rely on your wits and your reflexes?

The answers to these questions are not merely game preferences, they are life choices. Different perspectives and desires lead not only to different styles of gaming entertainment, they also lead to different careers, different relationships and different lives. This is how video games are like personality tests. When you tell me what you like to play, you're telling me who you are.

THE PIONEERS: WHAT IS A VIDEO GAME PROGRAMMER?

If you want to pioneer a new medium, you've got to have pioneers.

But what does a video game pioneer look like? Do I want a technologist, gamer, entertainer or joker? Perhaps a hybrid of some or all of them. There are many ways to think about video games. For me it was an exercise in communication, a piece of broadcast media. Then again, it's also a biofeedback loop.

Though all this is true, there's no getting around the fact that, first and foremost, a video game is a computer program. Whatever else one might say about game engineers, they must be programmers.

Remember, a computer program is a set of absurdly simple instructions combined together in a way that is far too complex for the programmer to understand. This is why computer programs have bugs.

[NOTE to the Non-Nerd: Also remember, a Bug is a logical malfunction where the computer is doing exactly what the programmer told it to do instead of what the programmer wanted it to do. If bugs don't fully crash the system, programmers attempt to reclassify them as "features" (this is nerd-speak for an unintended/unforeseen consequence I do not want to fix).]

But Video Game Engineers aren't just programmers, they are a special breed, particularly back at the dawning of the industry. If you'll bare with me, I shall attempt to expose the naked truth about early video game pioneers. Spoiler alert: It's all about special cases…

There are two kinds of people in the world: Those who divide the world into two kinds of people and those who don't. I frequently do.

One of my favorite kinds of two kinds of people is Splitters and Groupers. Groups are supremely important to us as human beings. From an evolutionary standpoint, our very survival depends on this idea. A naked human alone in the wild is extremely vulnerable. But a group of humans, united and making tools, becomes the apex bio-competitor. Deep down in our DNA, we know that groups are an essential key to our safety and survival in the world.

Groups are all about similarities and differences. Similarities help us separate members from nonmembers. Differences help us separate members from each other, thereby splitting the group into subgroups.

I'm fascinated by patterns of similarities and differences. That's how I make sense of the world around me, by sorting out what's what and then choosing which to engage and which to avoid.

Splitters and Groupers is an interesting way to see things, because every time we declare a group we create a new split, and every time we declare a split we create new groups. Splitters and Groupers are truly the Yin and Yang of two kinds of people.

That's interesting, but what's the point? Well, this profoundly pedantic preliminary preponderance proves an apropos precursor to my proximate pending postulate. That being:

There are two kinds of programmers. Over time I've developed a theory about how most programmers fit into one of two categories. I call it the 80/20 Model of Programmers. In my experience, approximately 80% of programmers prefer objective facts, unambiguous figures and indisputable results. They love that computers do precisely what you tell them, behave consistently and never lie or play politics. The other 20% are more subjectively focused and enjoy expressing themselves. They are always looking for an outlet and see the computer as a vast frontier, teeming with previously unknown creative possibilities. Infuse a nerd with ham and you've got a 20%er.

80%ers tend to see their computer as a refreshing alternative to dealing with people. If they weren't programmers, they might be accountants, fact checkers or actuaries.

On the other hand, 20%ers tend to see their computer as an intricate and complex conduit, offering new ways to connect with people. If they weren't programmers, they would likely be artists or performers of one sort or another (and probably starving).

For more than a quarter century I hopped all around the software industry. I've been an employee or contractor for more than two dozen companies doing all kinds of software development. In all that time, this perception (and proportion) has remained solid. Everywhere I worked there were mostly 80%ers and some 20%ers… with one glaring exception: The video game industry uniformly violates the 80/20 model. Game programmers are almost exclusively 20%ers, and Atari was populated by some of the most perverse 20%ers.

This isn't a coincidence. 80%ers have a hard time with games because they don't tolerate ambiguity well. Engineers in general (and 80%ers in particular) like things clearly defined and well thought out. Most software just needs to meet the technical specification, which is an objective criterion. No one says, "This word processor isn't nearly as exciting as my C++ compiler." If text box A pops up when I click button B, I know my program works and I know I'm done. This is perfect for 80%ers.

A video game is different. A video game must still meet all its technical specifications, but it also has to be fun. This is a very subjective criterion. When you take people who spend their life chronically avoiding ambiguity and ask them to add "It's got to be fun" to their technical specification, well... What does fun mean? How do you define fun? How do you measure it? Can you describe in clear, unambiguous detail, a process which will guarantee fun? (let me save you some time here, the answer is no)

You spend up to a year of your life pouring your heart and soul into the code, crafting and polishing each aspect and nuance from every angle you can imagine. All those days, nights and weekends lead up to that one special moment when you go sit behind the two-way mirror and watch your work presented to the "client." And who sits in judgment of your work? Some eleven-year-old boy, getting paid a pittance and pizza to express his opinion. After two minutes he tosses the controller and issues his verdict: "This sucks!"

And you know what? He's right! How do you deal with that? Egoless programming? I don't think so. 80%ers don't like that at all. Though passionate about playing games, they'll leave the making to others.

Video game development is different from typical programming. It requires a different approach. You have to put yourself on the line. Good games don't come from disengaged people, they come from neurotic oddballs who seek out this kind of emotional punishment. Why would anyone do this? Because "the entertainment industry." Because 20%ers need to have their work seen and appreciated. Because when someone plays *your* game and loves it... that feels amazing! It's unbelievably gratifying!

But you can't get there without personal investment, which means taking some risk. An interesting thing about classic game developers is many of them really enjoy casino gambling. We used to have occasional scumbagathons. A scumbagathon occurs when a number of developers jump in a car (or cars) and head for Reno, whose Casinos were a mere 3½ to 5 hours away (depending on who's driving). Every once in a while, some spontaneous stress relief was mandated, and we answered the call.

As a therapist, I can't help but point out that gambling is inherently stressful. Isn't it interesting that people who are outrageously stressed seek a stressful activity to destress? It may seem counterintuitive but consider this: People who seek high stress work situations frequently seek high stress leisure activities. After all, if people didn't have a predilection for stress, they wouldn't pursue this kind of work in the first place. Many believe the best cure for stress is placid relaxation. But for stress lovers/junkies, stress isn't the problem, it's the goal. What they're looking for is a change of pace, to avoid getting burned out on the same type of stress. You may love bon-bons, but after a heroic binge you might be sick of them. What do you do? Quit snacks entirely? More likely you just switch to pastries for a while. But don't worry, love endures. It won't be long before those bon-bons reappear in your shopping cart. Meanwhile...

Programming the VCS was a nerd's paradise. It was keeping lots of nitpicky pieces of data and complex interactions in your head, while having to organize and write procedures which must work at a precision of about a millionth of a second. For many, this is like drinking a pain & drudgery cocktail. But for some people (nerds like us), it's a delightfully intricate puzzle to solve. We thrive on the challenge.

The kind of programming we did at Atari was "hacking". It was not formal, well-structured code with well-documented algorithms (like I was taught in grad school). It was down-and-dirty getting things done in the cheapest way possible. This means both wonderfully elegant and brutally inelegant approaches, intertwined together like spaghetti. When it works, it's delicious! It also means doing things that professional programmers should never ever do. As a formally trained software engineer, I was keenly aware of the crimes against programming standards one had to commit to be effective on the Atari VCS. *That's* what I'd been missing. All the joy I'd discovered in school, then lost at Hewlett-Packard, I found again at Atari.

This kind of programming is way more fun than formally structured software development. The goal of most industrial programming is to make things maintainable since they will be constantly modified and enhanced. But at Atari, I make the game once and it will never be updated. It just needs to work, and I don't care how. Get it right, get it done and get it out. There's a beauty to that. The hacker is a very romantic image. But it's not what software development was supposed to look like at the time. In fact, it was the opposite. We were guerilla programmers, forging a new frontier in the unspoiled, untamed software wilderness.

Some guerillas had graduate degrees and others were self-taught dropouts. Some learned at school, some learned at work and some learned at home. It didn't matter how you got it, it mattered what you did with it. The VCS technology was so primitive, you had to be a hacker. You had to do bad things to produce good results. So many ugly decisions and trade-offs. It was a tense high wire balancing act. And if programming obstacles aren't enough, it has to be entertaining on top of just plain working. That's a *huge* challenge.

Making video games at Atari required a rare mixture of technological prowess and creative flair. You must be nerd enough to master the machine, and artist enough to do something worthy of a player's attention. You had to be a hybrid, because the VCS forced you to exploit both skillsets. This mixture of tech and art was eye-opening for me. Atari was the first place I found it, and found I needed it. After Atari, it would take nearly thirty years to find it again, and I'd have to become a psychotherapist to do it.

[NOTE to Most People: If you are flummoxed by my transition from programmer to therapist, it's possible you think of all programmers as 80%ers. I'm a solid 20%er. Whereas 80%ers get into computers to avoid people, I ultimately got into people to avoid computers.]

At Atari, a game was a work of authorship as opposed to a collaborative effort. In collaborative projects, the artists, designers and programmers all do their thing, coordinated by management. But an Atari game engineer *is* the artist, designer, programmer and project manager. On top of that, you need the discipline and focus to drive the project all the way through to a timely completion.

Atari was looking for self-motivated innovators, uniquely skilled people with passion and unusual perspectives. They were looking for me. For the first time, I'd found my calling. Making video games at Atari was everything I needed my life to be.

So, what makes a good game programmer? Someone who's both anal and silly. A detail focused wide-eyed dreamer. A goofy visionary, on a quest to bring their vision to life by conjuring a magic blend of elegant simplicity and bold innovation (the Yin and Yang of game design). Not a bad approach to life, either.

We were changing the world and we knew it! We didn't have any idea what that meant, but we loved every minute of it because we were doing it with a fascinating group of people. Each programmer had some other significant talent or skill, no one was unidimensional. And since we differed widely in experience and temperament, interesting thoughts and concepts emerged as we bounced off the walls and off each other. Some ideas were unstable, some downright toxic, but some were beauty incarnate. It was an incredibly productive creative environment… initially.

Now that you have more of an understanding of game engineers and their technology, I'd like to share with you another huge problem which contributed to the great crash.

Atari wasn't all chocolates and roses, there were plenty of thorns. Some of the prickliest came down to corporate cultural differences, and the communication errors created by those differences. No place was this difference more apparent than the relationship between engineering and marketing. Sadly, one of the real casualties of the cultural transition from Nolan to Ray.

Game programmers are an odd assortment of talented eccentric people.

Nolan understood this and insulated engineering from the rest of the company. Ray's regime wanted engineering to conform more to standard corporate expectations, resulting in a situation best described by this old saying from rehab circles:

An expectation is a down payment on a resentment.

CHAPTER 9
CONFLICTING FEELINGS

There are two kinds of people in business; people who make things and people who make money from people who make things.

People who make things just want to make things. People who make money from people who make things just want to move money from other people's pockets to their own.

[NOTE to the Non-Bifurcator: The Visionary is the third kind of this particular two kinds of people. The visionary is an interesting hybrid, being simultaneously both and neither of the first two. Visionaries supply inspiration, goals & challenges to the people who make things and the people who make money from people who make things. They also supply anxiety at no extra charge.]

In engineering the product is the goal. In marketing the sale is the goal, the product is merely the vehicle. It's a very different mentality, and a potentially volatile mix.

There was one night in particular which really highlighted this difference to me, the difference between creators and merchants…

It was a warm July night in 1982. The E.T. "situation" had yet to impose itself upon me. Joyfully cruising down the road with the T-tops off, I felt the warm summer air blow through my locks as they flailed about on the now deserted dancefloor that is my scalp. I was driving to a friend's house, a place where many enjoyable evenings had been spent. I was driving particularly carefully. On my previous visit I was pulled over and ticketed for speeding. Ironically, it wasn't speed I was really at risk for, it was the coke in the glove compartment. Fortunately, that never became a topic with the officer. Tonight's party favors are quite different. Though not an illegal substance per se, they do constitute significant legal exposure.

I'm en route to a clandestine meeting. If word of this evening's events reaches any of several executive offices, it is likely that people will be fired, millions will be spent in legal fees and careers will be ruined... all in pursuit of correcting a wrong that will never have taken place. But none of that will happen because none of the participants will ever talk about it. Until now, that is.

Some colleagues and I are gathering for the specific purpose of sharing our yet-to-be-released games. It's a show-and-tell among friends. However, since we represent more than one video game company, it's understandable how some might see it as industrial espionage, so we keep it on the down low.

It's a strange thing when people you care about leave to become competitors. Some feel abandoned and embittered when this happens, but tonight's crew remained friends. We continued to hang out together, party together and root for each other. It's business as usual, except what used to be casual conversation is now potential infringement of trade secrets. So that was different. And to what end? Competition isn't the same thing to creators that it is to merchants.

One of the best things about working in vid games back then was reviewing each other's work, exchanging ideas and increasing the world's fun supply. But now we are legally barred from doing it. In showing our work to our marketplace adversaries, trade secrets could be stolen, and competitive advantages lost. We understood this line of thinking and it makes a lot of sense. In *this* circle however, no one is going to steal anything. We are smart enough to understand what we're seeing, earnest enough not to abuse that knowledge, and we all have enough ego and mutual respect to want to show our friends what we've done.

So, we do it. We get together and show each other our work. Violating every cannon of safe business practice and invalidating millions of dollars of previous (and ongoing) litigation.

To add a twist to the evening, we decide to keep authorship a secret at first. That's *our* game, guessing who did what. Remember when I told you how everyone signs their work?

Everyone is unique, and early video game programmers even more so. As each game appears on screen, I see unmistakable traces of personality.

Our styles and quirks shine brightly in our games. I was able to identify every author that evening because the same traits I enjoy in the people tend to show up in their work. These varied and eccentric personalities account for the vast array of entertainment created for the VCS. They produced a far more extensive offering than most thought possible for such limited hardware. They also provided one of the most compelling work environments ever.

We spent the evening sharing our games and techniques, but mainly just enjoying our friendship. And that's the story of the secret meeting we never got sued for. If any people who make money from people who make things had heard about this meeting, it could have gotten very ugly. Fortunately, none of the potential legal pitfalls were ever realized. Nothing was stolen and no product releases were compromised. It was simply some people who make things, having innocent (though dubious) good times.

MARKETING & ENGINEERING: WHAT'S THE DIFFERENCE?

What engineers hate more than anything is to be asked to violate the laws of physics. And this is something that marketeers routinely do.
NOLAN BUSHNELL, ATARI CO-FOUNDER

Without conflict there is no drama.
MANY SCREENWRITING GURUS

Marketing's goal was creating and expanding market channels to target and sell through. Engineering's goal was making good games to fill those channels. It makes sense these goals would work well together. However, when you expect things to make sense, you're losing touch with Atari.

Both have appropriate goals: best product and best sales. Each feels they are on a mission to accomplish something extraordinary. This can be a basis for historic achievements. Initially things were fine, but over time the conflict between Atari marketing and engineering grew to legendary proportions. Each came to see the other as a liability. Instead of supporting each other, we lost respect for one another.

It's like two kids fighting over a game they're trying to play. Each thinks the other is the problem, but the truth is they're both trying to achieve the same thing, they just can't see how to do it together.

When you're playing a game that involves people's careers, their egos, a global spotlight and billions of dollars, it's very intense. Things can get out of hand. That's when a therapist comes in handy.

The issues that evolved between engineering and marketing were a product of who we were. Personality clashes and cultural differences led to a divergence of paths. Instead of growing together, we went tribal and splintered. This helped no one.

THEY DON'T THINK LIKE WE DO

Are engineers and marketeers really *so* different? If you prick us, do we not bleed? If you praise us, do we not seek raises? If you lay us off, shall we not file for unemployment? Clearly we had similarities, but our priorities and values were as different as our attire.

Marketing wants things to be meticulously organized because predictability has real bottom-line value. Engineering wants things to be chaotically organized because this presents greater opportunity for breakthrough concepts. It seemed at times like marketing was willing to sacrifice potential for reliability. Engineering saw this as innovation suicide, the antithesis of their ideals.

For instance, there was the time when a memo from marketing went up on the bulletin board. It expressed concern that engineering was unproductive from marketing's point of view. They cited "unrestrained creativity" as a problem in our department. The implication being that we were undisciplined in our execution and not sufficiently meeting their demands. On some level it felt like they were saying "Stop screwing around. You should produce games as fast as we buy licenses. Hurry up or we'll miss the window!"

This would be fair if we were screwing around, but we weren't. We were working to create innovative concepts and approaches on an increasingly outdated hardware. Engineers believe inferior product shuts windows and locks doors, rendering licenses useless.

This is the essence of the battle between license-power and product-quality. Engineering wants good games. Marketing wants saleable products. These are not necessarily the same thing.

As you may imagine, the unrestrained creativity memo did little to inspire us to make better games. It did foster some resentment though.

Another significant point of divergence is tolerance for ambiguity. Engineers tend to have low tolerance for ambiguity. Marketeers and management, not so much. This is largely because the world of engineering is based on accuracy. If I'm fuzzy with my bridge or building specs, people can die. Therefore, it attracts those who prefer an objective environment where specificity is valued.

On the other hand, the world of sales and marketing is based on perception. Since impressions play a dominant role, this world tends to attract people more comfortable with shades of gray. It's a subjective environment where fluid adaptability rules.

Ambiguity consternates engineering efforts, whereas marketeers may rely on (or even foster) ambiguity to achieve their goals. There is no right or wrong here, but there are consequences…

When marketing sells an illusion by promising the impossible, engineers take offense because this compromises their integrity. After all, marketing is writing checks that engineering must cash. But marketeers compete with other puffers, and if we don't get in the door and make sales, we can't pay wages or produce products.

In the big picture, the difference between engineering and marketing is that engineers tend to under-promise and over-deliver while marketeers tend to over-promise and under-deliver. We're built to step on each other's toes when we try to walk together.

THEY DON'T UNDERSTAND WHAT WE DO

As Rob Fulop was finishing his first project at Atari, someone from marketing saw it and was quite impressed. "Wow, that's an amazing effect! Where did you get the idea for this?"

"It's Night Driver," Rob replied, "it's one of our best-selling arcade games." Rob was porting Night Driver to the home system, and it was a remarkably faithful reproduction. Rob was a bit shocked this marketeer didn't recognize one of our own top products.

Marketeers' comments were rarely well received, because they would look at the games but almost never play them. They seemed oblivious to how gameplay figured into a game's success. Consequently, they had no contribution to that aspect of the games, which was a shame since that's where engineering was most receptive.

Instead, they requested impossible things, the classic was, "Can't you make the ball round?" Well, maybe, if we eliminate the player…

Marketing repeatedly asked for things that can't happen, but rarely offered suggestions for things that could. I never expected them to understand the technology but playing the games does not seem like too much to ask. It felt like they were disconnected from the product they were selling.

One place this disconnect shone brightly was consumer testing. Marketing wanted to figure out what people wanted in a game, but they didn't seem interested enough to play the games, understand the experience and judge for themselves. They relied heavily on consumer testing to make up the difference.

You may recall the testing ordeal I endured with Yars' Revenge. Though I recounted many lessons learned back then, I saved one of the biggies for this moment. It revolves around another major sticking point from the dawning of video games…

GETTING WOMEN INVOLVED IN GAMES

The world was playing (and paying) video games. Well, most of the world. It was clear from the start that women were underrepresented in the player population. To engineers, this was a lot of joy potential being lost. To marketeers, this was a lot of money left on the table. To management, this was a problem to solve so let's get engineering and marketing to fix it. Getting women to play video games was like the weather, everyone talked about it, but no one did anything about it. Well, almost no one.

[NOTE to the Now-vs-Then Stat Freak: Today, women are roughly 45% of gamers. This is up from 35% around 2000, but back in the early '80s it was just 20%. However, while women were just one in five arcade players, they were more than 50% of the Pac-Man players. This fact greatly interested marketing & management.]

To be fair, women were also underrepresented in *making* games. Which maybe isn't so fair, but I can honestly say that most early video game makers were exceptional people, and the women making games were no exception.

Carol Shaw created some of the most innovative and challenging software at Atari, then went on to do River Raid at Activision. Carol left Atari shortly before I arrived, but her legacy was profound. My colleagues frequently spoke of her programming prowess.

Dona Bailey was a talented programmer who became the only woman working in coin-op engineering. She co-created Centipede, one of my personal favorites. Centipede was notable in that it was quite popular with women… which was no coincidence. Dona worked with Ed Logg specifically to produce a game with broader appeal. Though we were contemporaries at Atari, I didn't spend much time in coin-op and never got to know Dona. My loss.

Another woman worthy of note in video game history is Marilyn Churchill (then Marilyn Thaurer). Marilyn was the first person hired exclusively as a video game artist. She went on to become the industry's first Art Director. From the moment she arrived, every Atari home game became more beautiful. She also brought Jerome Domurat to Atari, and I'm eternally grateful for that.

Then there was Carla Meninsky. Carla and I shared similar paths to our current job: Raised in metropolitan New York suburbia with the traditional goals of going to college and entering corporate business life, now reveling in how Atari obliterated that vision. We were kindred spirits, who left corporate monotony for Atari just as children bust out of class for recess. We were both high-spirited, but Carla had a far greater sense of subtlety and propriety than did I.

Carla created Dodge 'Em, Warlords and Star Raiders for the VCS, three excellent and successful titles. Naturally, she is a logical choice for conversations on the topic of bringing more women into the video game market. And it comes to pass that Ray Kassar, during one of his wild safari adventures through engineering, steps into Carla's office to consult the oracle and ask the fateful question.

Ray comes prepared with a few suggestions of his own, which he offers after a perfunctory exchange of pleasantries, "Can you do something like 'Let's go shopping', or a game where you design your dream house?" Carla is dumbstruck for a moment, then recovers and says, "No, Ray. That's not the way you get women involved." His goal is reasonable, but his approach seems a bit antiquated.

Marketing continues to ask: "How can we turn more women into gamers?" Solutions, however, remain elusive. But on at least one occasion, an unexpected answer presented itself. It's another in the series of twisted ironies that *is* life at Atari.

In 1981, Pac-Man is all the rage. One thing about that game catches every marketeer's eye: Women love Pac-Man. No one knows why, but something about it works for them, that's for sure. Meanwhile, Yars' Revenge is becoming the most evaluated video game in Atari history. Here is the result I didn't tell you, and where it led.

When the smoke cleared after the big event, Yars set the record for high score in a play test. That's nice, but the really interesting news is the demographic breakdown. Of the various subgroups, the one that rated it highest was adult women. Adult Women!

On one hand, I feel a profound sense of pride and accomplishment since appealing to adult women is always a priority of mine.

On the other hand… Hello? Marketing?!?!?

This becomes the core topic of many conversations with product managers, but only because I keep bringing it up. We talk about how they want a game for women, and this is obviously a game for women, so how are they planning to market it to women? The answer always comes back the same:

"Oh, we're not going to market it to women." Why not?

"It's a space action shooter and women don't like space action shooters." But your own testing says they like this one!

"Trust me, they don't." Then why do you test the game?

"To find out if people like it."

Aaaaagggghhhhhhhh! I'm trapped in a Dilbert cartoon before they exist. I continue our chat longer than I should. Once a salesperson says "trust me" the conversation is pretty much over, but it's hard for me to disengage from this kind of absurdity… which isn't good for me. In fact, I believe years of talks like these are partly responsible for my becoming a psychotherapist.

The idea their test results don't seem to mean anything fuels another issue as well: It fosters an unhealthy disrespect on our part by creating the impression they don't know what they're doing.

Let's face it, it's easy to cast marketing as ridiculous for not listening to their own data. But there you go expecting things to make sense again. And it's not like they never listened…

Sometimes they would grab isolated results from a number of separate focus tests and assemble the "Frankenstein spec" for the perfect game. That monster frequently bit them in the butt. With little sense of gameplay, they couldn't suggest how to combine those pieces into a coherent game concept. Neither could we. Part of the problem may be the player responses. People can tell you if they like a game or not, but they're not terribly credible when you ask them why.

As it happened, the testing became moot in short order. In fact, it pretty much stopped after the whole Yars' Revenge debacle. It felt like they'd arrived at the belief there is no point in assessing game quality because any game we release will sell, as long as it has a license attached and gets out the door on time. After all, this was their experience so far.

Nonetheless, the big question remained: What draws women to video games?

Although I was not particularly enamored of marketing, I was no less curious about the answer. Now there were two data points: Pac-Man & Yars' Revenge. These are two vastly different games, but after thorough and careful analysis I gleaned one salient commonality between the two: They both rely heavily on an oral component. In Yars you eat away the shield to obtain the big weapon. In Pac-Man, eating is how you relate to the entire world.

It appears the key to permeating the great video gaming gender barrier is incorporating aspects of conspicuous oral consumption into the game play mechanics. But that's a bit of a mouthful. Perhaps it's easier to swallow when posed thusly: How do you build a video game with universal appeal? Bite by bite. So chew on that for a bit.

AREN'T WE ON THE SAME TEAM?

Marketing seemed completely disconnected from what it took to make the product. After months of abandoning our lives to pour everything into conjuring tech-magic on the screen, it was hard to hear a litany of complaints about what wasn't there, and so little appreciation for what was. Imagining things is easier than programming them.

We felt underappreciated, I'm sure they did too. We knew they didn't understand our challenges. It was less clear how little we understood theirs (less clear to *us* at least). Presumptions about each other undermined our ability to resolve these differences and improve our product and process. It's hard to find people who can cross conceptual chasms and bring both sides to the table in any negotiation. At Atari, it was impossible.

My good friend Jerome took it as his mission to educate people about the process of making a video game. He did this every chance he got, for anyone who would listen. When marketing and management made their way to the graphics lab on dog & pony days, Jerome would regale them with a multi-media demonstration of how animations were composed and inserted into a game. His audience would always be fascinated because he did it so wonderfully. Jerome grew up making hundreds of dioramas, beautiful elaborate dioramas. He told me once how he'd set up a "show" in his home and charged friends a nominal fee for the tour. It's no wonder he wound up working in some of the nation's most prominent museums, creating engaging displays for the public. By the time he found his way to Atari, Jerome was quite adept at creating dynamic presentations to convey information. If every Atari employee had been required to attend a few Jerome "showings", the company would have been far better off.

JEROME IN HIS SPACE.

PHOTO BY:
DAVE STAUGAS

Alas, the vast majority of marketeers and managers never saw a single one. They didn't know much about development, nor did they care to. By mid-1982, it seemed VCS development had become a strategic afterthought, relegated to filling orders. We were fed a steady diet of licenses and completion dates. This was a bitter pill to swallow. We were constantly coughing it up and spitting it back out at them. This style of management sowed unnecessary discontent and led to some horrendous business decisions.

Consider the fact that one of the biggest game properties of all time was given the shortest schedule... by a factor of five! This demonstrates how little insight there was into development realities and how little importance engineering held in the eyes of management. Believe me when I say we will talk more about this down the road. Bob knows it's true.

But first, let's finish with marketing and engineering. That relationship played a critical role in Atari's success and failure. As in any relationship, one sure indicator of a rotting foundation is when brainstorming gives way to blamestorming. Sadly, Atari had practically cornered the market on blamestorming talent.

Marketeers came to see engineers as pedantic idealists mired in detail, while engineers came to see marketeers as superficial manipulators. We were a bunch of entitled spoiled kids running free, with no discipline or supervision. They were a bunch of uncreative automatons who "brought value" by offering critiques which served only to affirm their cluelessness. These were the stereotypes. In truth, we were simply doing our jobs, which just happened to violate the other's sensibilities. No one wanted to admit what they didn't know, which was a big problem for everybody. It got to the point of ridiculocity.

We were all working toward the same goal, but we weren't working together. It wasn't all bad. At one point in late 1981, Jewel Savadelis from marketing was made interim head of the VCS department. Jewel was a gem, and her stewardship ushered in a golden age of interdepartmental harmony, providing a healthy and solid connection between the two departments. This lasted until she inevitably returned to marketing. Thereafter the connection faded, only to be replaced by incidents like the unrestrained creativity memo which poisoned the well and deepened the cultural abyss between us.

It became a competition among very smart, aggressive people, each with their own agenda of how to move the company forward for everyone's benefit. Each came to believe: "I'm trying to do the best thing for everybody and they're opposing me. WTF?" Now we're deep into BMOBS territory. This is a very destructive mindset, eating away at the soul of the company.

The competition raged on. Hanging in the balance was the power to make decisions, including one of the most important decisions of all: Deciding when a game is ready for release.

IS IT DONE YET?

In the beginning, engineers made games. The goal was simply to make something fun. A game was done when the game makers (as a group) decided it was good enough. Worthy games were handed off to marketing and manufacturing where they were packaged, reproduced and distributed for sale. Sucky games went back for rework. Engineering did gameplay, and marketing did theme and titling.

As time went on, learning curves kicked in. As engineers gained experience and discovered more tricks with the hardware, the quality of the games increased significantly. Compare first wave games from 1978 or 1979 with the games of 1982. The difference is dramatic, but the game maker's goal remained the same: Create quality games.

Marketeers underwent a learning curve of their own. The innovations they contributed were licensing and release schedules. Originally, they were selling what we developed and throwing in the occasional license for a hot coin-op game *that seemed feasible* on the VCS. In time they branched out to any profitable coin-op game and ultimately to non-game properties, specifically big-ticket Hollywood movies and television shows. They were all about pre-sold market concepts. This makes sense and sounds like good marketing, but increasingly their priority seemed more about hitting the pre-sell window and less about delivering a solid game.

Engineering wants to forge a solid product. Marketing wants to strike while the iron is hot. As the overlap shrinks, this divergence becomes a source of frustration and conflict… for everybody!

Competing corporate agendas are nothing new, but they force the question: Who has the power to make their agenda *the* agenda? I had arrived at Atari just as this battle was coming into full swing. The power to control releases was shifting from engineering to marketing, and that was a big shift.

This is what Alligator-clip Jim was talking about in the cafeteria on my first day. When he said, "...it isn't what it used to be. This place used to be amazing," he was ruing the early signs of the transition from quality-based to schedule-based criterion for releases. By the end of my first year I was beginning to see his point.

How lucky was I to arrive at Atari when I did? Had it been a little later, my proposal to sidestep Star Castle in favor of a nameless "better concept" could not have succeeded. An assigned coin-op conversion would never be allowed to shift to something else, because the whole point of making the game would be to exploit the coin-op's audience, regardless of game quality. Dennis would have to say, "Sorry, we need to do the license and it has to be done by..." And poof, Yars' Revenge would never have existed.

This also raises an interesting question: Is quality important? For short term sales on pre-sold properties, maybe not so much. But for longer term concerns like corporate image and customer loyalty, I think it is crucial. Many cite a sudden glut of low-quality games rushed to market as a seminal reason for the video game crash.

But this is before anyone sees a crash coming, and the interdepartmental struggle is ongoing, largely because we can't see the inevitable conclusion. Spoiler alert: Marketing is destined to win, due to a phenomenon I call Warshaw's Law of Marketing Inversion.

Consider this: Most companies start out engineering heavy, because there isn't much to do until they have a product. Since engineering represents most of the early budget, they receive a great deal of management's attention (especially if senior management contains engineers/developers). However, if the company starts to succeed, sales and marketing may scale up rapidly to meet the growing demand, while engineering continues making product.

Warshaw's Law of Marketing Inversion states that in a successful company, engineering resources grow arithmetically while sales and marketing resources grow geometrically. The resulting budget imbalances inflate the visibility of marketing in the eyes of management, while diminishing engineering. This tends to shift the power to define corporate direction in favor of marketing. Simply put, the expensive wheel gets the grease.

Ultimately, engineering has little hope of dominating corporate mindset. For Atari, it's easy to spot the precise moment this transition really began to take hold. The moment Nolan and the other founders signed papers to cede control to Warner, that's when it all started to change. A massive corporation which knew nothing about technology was very excited to bring their brand of corporate-think to a fledgling techno-tainment enterprise. Ray Kassar was just the delivery system, the direction and outcome were already assured.

WHAT'S THE DIFFERENCE?

One quintessential point of contention which highlights all the nuances is the debate between Make-It-Better and Sell-It-Now.

It's the classic struggle: Product-pull versus Sales-push mentality. Marketing wants to sell it now because we're losing critical sales every day we delay. Engineering wants to wait and improve it because better products sell more units. But how many more? What's the difference between what a product might sell later and what it will sell now? At Atari, the answer is lots of antacids.

Today, the internet solves this by allowing developers to update a product after release. But in the early '80s there was no internet.

During 1981 and 1982, the sales-push mindset grew steadily, taking over Atari. It was about getting a recognizable property, pumping it with advertising, reaping quick sales and… that's it. There seemed to be no consideration that product quality might impact our reputation, future sales, etc. It was all about hitting the window while it was open. And E.T. was the crowning achievement of this mindset.

As a developer I believe better products do better, and as an entrepreneur I realize this increment can be impossible to measure (or justify). True, a week lost to delaying release can never be recovered, but losing sales because a product sucks doesn't help either.

Differences in ambiguity tolerance, making things or money, tendency to over/under promise, etc., these are not just opinions, these are cultural stances. They may sound theoretical, but they manifest in painfully real ways. For instance, one company decided to make a product for their most expensive license ever... in just 5 weeks!

CHAPTER 10
DO IT IN 5 WEEKS

VOYAGE OF DISCOVERY

Developing a video game is a Voyage of Discovery, which is a poetic way of saying: I'm not sure where I'm going but I'll know it when I get there. A voyage of discovery happens because I'm trying to do something innovative, so I'm starting with a few vague ideas and lots of blanks to fill in. If I had a more detailed plan it wouldn't be a voyage of discovery, it would become an excursion to a known destination. Excursions take far less time.

Normally, making a video game on the VCS system is a 1,000+ hour marathon spread out over 6+ months. This is the essential voyage of discovery, meandering about while remaining committed to a basic vision over the long haul. The plan evolves as new concepts and capabilities emerge. By now I've completed two such marathons successfully for Atari. Love it.

This schedule, however, only allows for a sprint. This means arriving at a known destination as fast as possible. This requires two things: a clear goal and an efficient approach. A thorough design creates the clear goal. Speed, however, is all about the approach. That's fine with me, when it comes to accelerating things, I'm already there.

In college, my girlfriend and I once took a math class together. On every exam, I was the first one finished and she was the last. We consistently tied for high score. (Interesting couple, eh? It's easy to see where we connect and where we conflict.)

When I decided to become a video producer, I enrolled in an 18-month certification course. I completed the program in 9 months and got my final project aired on PBS (a first for any student).

In fact, my initial college career was a sprint. Here was my approach: I looked deep inside myself to honestly answer the question: How long can I be in college? The answer came back: Four years and *no more!* After four years I will turn into a pumpkin. Fine. I'll just plan to make the most of that time. I didn't start with an educational goal. I started with a time limitation and sought to maximize my results within that constraint. Wait, I feel a digression coming on...

There are two kinds of people in the world: Maximizers & Minimizer's. We maximize pleasure or minimize pain. We maximize potential or minimize risk. Maximizers start insurance companies, minimizers buy policies. Though we can (and often do) switch between them, most people tend to lean more one way than the other. There are significant personality differences between maximizers and minimizers, but I'm saving that for another book. Meanwhile, in this book...

Some students minimize study time or class time, I set out to maximize my return on a four-year investment of my time.

College is a system. As with any system, you figure out how it works and then you try to hack the system to exploit it for some advantage in cost, time, results or whatever you desire. That's what systems engineers do. I was acting like a systems engineer before I even knew what an engineer was. I studied Tulane's "manual" extensively to figure out what might be hackable. Then I created my plan.

Considering I had no previous experience (or desire for) being a good student, the first thing I needed was a better perspective. With a solid framework to guide my decisions, I might become a decent student. Some see college as an academic pursuit, others see it as a party break before facing adulthood. Neither of these worked for me.

After giving it some thought, I decided to view college as a project. My deliverable was a useful degree. I accepted my four-year schedule limit and set out to do my maximizer thing. That's not how I phrased it at the time, but that's how I acted.

So how did college go? I've alluded to the fact that it went pretty well. Here's the first few paragraphs of a book I wrote about it, reprinted here by virtue of permissions granted to me by myself:

> *I conquered college.*
>
> *By this I mean: I received my Bachelor of Science degree from Tulane University (in New Orleans) with honors, sporting a cumulative GPA of 3.9. I completed a double major in Economics and Mathematics, with an additional minor in Theater Arts. You will see evidence of all three throughout this book.*
>
> *I made the Dean's list every term and was recognized as a Tulane Scholar. I was inducted into the freshman honor society, the math honor society, and Phi Beta Kappa, the national honor society.*
>
> *I did all this in three years - not four - saving lots of money.*
>
> *On the strength of this record, I was granted a full scholarship to the school of engineering where I completed my Master of Computer Engineering degree in one year (GPA 3.9).*
>
> HOWARD SCOTT WARSHAW, CONQUERING COLLEGE, 1992

Fortunately, the idea of working within a fixed schedule is not new to me, so the planning for E.T. comes pretty naturally. I swap "five weeks" for "four years" and do my maximizer thing. I've spent 18 months studying the VCS system, so I'm comfortable with it and already have some possible hacks in mind.

And there's one other thing: Despite this obscenely short schedule I still want to do a breakthrough game. This has to be a sprint, but I still want marathon results… and I believe I can do it. I'm not sure exactly what I'm full of at this point, but whatever it is, I am absolutely overflowing with it now.

Sprinting a marathon. I've never done that before. How do you train for sprinting a marathon? It doesn't matter, because there's no time for training. Doing it will be the training.

If I'm going to pull this off, I need a plan. An extremely detailed plan of precisely what I'm trying to do. I need to be as clear as possible about the specifics of my final destination. Video games have a name for exactly this kind of plan: The Design. Fortunately, I've got one of those, and it's already approved.

DESIGN OF THE TIME

Over the last four decades I've done more than 100 interviews about this game. Invariably they ask some version of :

"What were you thinking when you made the E.T. video game?!?"

This is frequently followed by: "OMG, all those #%$&ing pits!"

But it's a fair question. What *was* I thinking? OK, let's pop the freshness seal on my brain and plug in a speaker…

For starters, I believe the fundamental key to success in any endeavor is knowing what you're trying to do and setting clear intentions. Each of my previous games had specific objectives for the design. For Yars' Revenge my goal was to create a game I enjoyed playing, as well as establishing myself as a capable video game maker. With Raiders of the Lost Ark my goal was to make the biggest adventure game anyone had ever seen on the VCS. There was no time component to those designs. I'll work on the game until it meets my quality standards, and that will take as long as it takes.

So, what was I trying to do with E.T.? Well, I certainly wasn't trying to make the worst game of all time…

Obviously, the big thing about E.T. is the little thing: the schedule. At a time when video games usually take at least 6 months to develop, I have 5 weeks to deliver a completed game for an extremely high-profile property.

Usually there's time to play around and experiment, but a schedule this tight doesn't afford much latitude. I have one shot to deliver some approximation of a credible game on schedule. I need a strategy to get the most results with the least risk. Looks like that Economics degree is going to come in handy after all.

My design objective for E.T. is simply to get it done. But how?

Most people say the secret to programming a game in 5 weeks is: Don't do it. It's impossible.

That won't work for me. But the other secret, the *real* secret, is to recognize that it's not a programming challenge, it's a design challenge. If I design a 6-month game and expect to do it in 5 weeks, game over. It typically takes over a thousand hours of work to deliver a good game, and it doesn't take a degree in math to see there aren't a thousand hours in 5 weeks, so that's not going to happen. I need to design a game that is doable in 5 weeks (technically 4 weeks and 5 days, since two-and-a-half days are taken up with creating and approving the design).

Ordinarily, you might think thirty-six hours for design creation plus a Learjet/limo roundtrip to Spielbergville is cutting things a bit thin, but realistically there isn't room for much more. To have any chance of delivering the game on time, coding must begin ASAP.

The first choice is genre. What kind of game will fit the bill? There are a number of basic game types; Combat, Racing, Action, Pattern, Puzzle and Sports. Combat, Racing, Action & Pattern games all share the same liability: tuning. They require striking a delicate balance of challenge and reward. This takes lots of time, which I do not have. Puzzle games need time for thinking up twists and clue planting, also time I do not have. And Sports games suck up mucho schedule time for artificial intelligence because the computer opponent must be worth playing. These types of games have too much complexity, too many moving parts. I need something different.

There's simply not enough time to do a big game, so don't think big; Think small. I need to remember the words of Leonardo Di Vinci: "Simplicity is the ultimate sophistication."

There is also a classic design mantra reverberating in my head, the KISS principle. KISS is an acronym for Keep It Simple, Stupid. This is a very important concept in programming, because simplifying things reduces the number of bugs and increases the reliability of time estimates. Simplicity should create fewer surprises.

Clearly, I need something simple. But it can't be too simple, or it will lack legs (replayability). It must be small enough to code quickly yet sophisticated enough to hold a player's interest. Simplicity is not enough; it needs to be elegant. Elegance is the key to creating a playable game quickly. It's also the challenge. How do I do it?

I shift my thinking to start focusing on the basics. At its essence, a game is simply a clear goal in a specific environment with a set of rules, obstacles and a well-defined start. What makes a game good? That's easy, a game is good if it's fun to play. Unfortunately, "fun" is difficult to define or predict, I just know it when I feel it. There are, however, some common themes that good games share.

In my opinion, one significant factor in a game's entertainment value is the relationship of game rules to game possibilities. The best games, the classic games, have the magic balance of few rules and a huge number of possible outcomes. Take Chess for instance, or Backgammon. And the quintessential model for thousands of years, the game of Go. They are elegant games, with simple rules yet vast horizons.

This formula is especially useful for video games because rules and environments cost programming time. Therefore, I need a game with few rules and a simple environment, which hopefully combine to create lots of possibilities.

How about a Treasure Hunt game? The basic treasure hunt game goes something like this: Find some treasure. Is this enough treasure? If not, get more treasure. When I have enough, use it to make progress or win the game. The game components are obvious: pieces of treasure, places to hide them, obstacles to increase the challenge, and some way to use the treasure to win the game. I need only tune the basic process of finding the treasure and using it. This sounds like something I can make happen in five weeks.

This will meet my limited schedule, but where is the replayability? I can shorten the dev time, but it still needs longer legs.

[NOTE to Pun-dits: Though it's true the E.T. character has short legs, that doesn't mean the game should follow in those footsteps.]

One key feature of a Treasure Hunt game is that players can actually win it. With the notable exception of Warren Robinett's Adventure, most video games of the era had no "win" condition. Players keep going until they run out of lives, trying to set new high scores. It's the How-High-Is-Up model of gaming. This is because most thinking about home game creation was based on the arcade paradigm. Having a win condition makes no sense in an arcade game because you need people to keep pumping quarters into the machine. But home gaming? Not so much.

So, how do you keep a "winnable" game fresh? Where are the legs? If the basic mechanics of the game are enjoyable enough (i.e. collecting and utilizing treasure), the player will want to do it again. I can randomly redistribute the treasure for another round, and as long as there are enough hiding places the challenge level should be maintained. Ideally, this results in an action puzzle that's fun to solve over and over again. I really wanted the game to have legs. In retrospect, it looks like I gave it enough legs to walk a short distance and fall into a pit.

Next question: How does all this fit with the movie? After all, at some point the video game will have to relate to the movie...

E.T. is a tricky movie for a video game adaptation. It's primarily an emotional tone movie, and the VCS is not exactly a facile canvas for emoting. And how will this movie set expectations for a gaming experience? Raiders of the Lost Ark is a profound action movie with a clear through line to translate, that's the stuff of which video games are made. But E.T. is all about feelings and wonderment. What does that video game look like?

When Steven Spielberg suggested a Pac-Man style game, I suspect he was thinking about the action sequence near the end of the movie. Elliott is biking around town (the maze) and people are chasing him (the ghosts), his friends are helping him along the way and E.T.'s in the basket as a wild card for special powers. This is a reasonable concept for a video game. I'd considered it initially but getting something like this programmed and playable at an acceptable level is unlikely to happen in 5 weeks.

Now let's think about the Treasure Hunt model in the context of the ET movie. It goes like this: ET assembles a phone from a variety of junk. Once assembled, he uses it to call home. Then he must evade interfering humans to meet and board the returning ship.

It's a beautiful multi-layered structure of quests. #1: Find all the phone pieces hidden in various places (pits). #2: Find the place to call home among all the "locations" in the world. #3: Find the designated landing zone in the forest (while avoiding the humans) in time to meet the ship. Three successive treasure hunts, each building on the last. As long as I have enough phone pieces, pits, locations and zones, the random redistribution each time hopefully creates a challenge fresh enough to motivate the player to try again. The player will be ET, this way they can have special powers more credibly, and it's probably a more interesting graphic than Elliott. This could actually work!

And there you have it. So, the next time anyone asks: "What were they thinking when they made that E.T. video game?"

That's what I was thinking.

SPRINTING A MARATHON

OK. I know what I'm trying to do. So... how will I do it?

This is one of my favorite things in life: An interesting problem to solve. I have a reasonable plan, but now it's all about execution. To make this happen, I'm going to have to answer the question: How can I be *most* productive over the entire 5 weeks?

The economist in me sees the productivity challenge as allocating myself most efficiently. Maximally! I enjoy racing against time (on *my* terms, that is). And, just like E.T., my brain serves me far better than my legs in this pursuit.

Since the speaker is still wired to my gray matter, I'll just start playing my rendition of: How to get more out of me. Please feel free to sing along or dance if the spirit moves you.

[NOTE to the Biomedical Engineer: Next time I want to hook up a speaker to my brain, please contact me and recommend Bluetooth.]

The first stop on my productivity tour is the age-old duality: Working Hard vs. Working Smart. I believe working hard is important, but the only thing working hard guarantees is fatigue. Working smart gets the job done. Working hard *and* working smart is the magic formula for big success. That's what I tried to do on E.T.

The next obvious question is one that didn't really come up: How about a team approach? Once people learned about the project (and deadline) this became moot, no one wanted anything to do with it.

Realistically, a group effort wasn't a viable option. A team software approach requires consistent application of formally structured programming techniques. As we've discussed previously, this is something the VCS will not tolerate. I love the misguided deeds one must commit to be effective on the VCS. However, the time needed to coordinate a team through all this complexity would chew up most of an already miniscule schedule, virtually assuring failure.

Then there's the too-many-cooks problem. I cannot afford to spend precious implementation time debating an already settled design (which would invariably happen). Hacking is a lonely business. It's hard to hack in teams without creating more problems than we solve.

Pursuing a group effort on E.T. is like asking nine women to have a baby in one month, it's not likely to happen. Yes, you can accomplish more with an army than you can with one soldier, but it's far faster to activate one commando for a quick-strike mission than to coordinate and launch an invasion. You may not agree, but I don't want to fight over military metaphors.

Do you remember the birthday party they threw for me on day 4? There was that poignant moment when I realized that doing this project will require removing myself from the best parts of my job. I'll have to work from home, long before it was fashionable. Our IT people set up a full development station in a bedroom and that became my home office. Now, no matter where I am when inspiration strikes, I'm less than two minutes away from putting it in the game. The only time that isn't true is while driving between home and Atari. Which doesn't mean I'm not working on the game, it just means it's a little longer before I can type in the changes.

Being maximally productive means bringing everything to bear on the task at hand. So why shouldn't sleep be part of the plan? When faced with a really tight schedule, some say: "If only I didn't need sleep, I could do this so much faster." Well, the reality is I do need sleep, but it doesn't mean sleep needs to be my enemy. I started thinking of how to use sleep as an ally.

Did you ever go to sleep with a problem only to wake up with an answer? That's dreaming up a solution. I would work until I was either exhausted or stuck on a bug, then I'd go to sleep. This way I'm looking forward to sleep as a potential helper rather than an impediment to my progress. It created a better environment for sleep which is a necessary part of the production.

Sometimes I wake up with a solution. I run to the dev station and type it in, then start playing with it and refining it and suddenly it's hours later and I remember I never finished sleeping and back I'd go. I never faced the anxiety of laying sleepless in bed thinking I should be working. After consistently grinding 16-20 hours per day, when I went to bed, I slept!

Sleep was my ally during the E.T. project, but it also produced one of the most frustrating moments of my entire game making career. One night I had a dream in which I was playing the best video game ever made. Everyone agreed it was perfect! I remember in the dream looking at it so carefully, analyzing it with the affirmative thought that I must remember every aspect of this game. When I awoke, I remembered the feel of the play and the sensation of the controllers on my hands. I remembered every detail of the dream except the particulars of the game itself. For months I tried all kinds of ways to remember, but to no avail. It was killing me. I've since come to the conclusion that the game was an illusion. The dream was only about the sensation of encountering the greatest game ever. At least that's how I dealt with the possibility of having had it and lost it, which can be a very hard thing to handle indeed.

As sleep is a necessity, so too is food. Most of my meals are eaten at the development station, which sports ample table space with easily cleaned surfaces. Sometimes, for a real treat, I go out to eat with a friend/colleague, usually Condon Freeman Brown. This way I can get away from working on the game for a little while and just enjoy some relaxing time talking about the game.

Condon is a manager in the VCS department, and for the duration of this project he is assigned as my keeper. He makes sure I eat and execute other activities of daily living. But Condon is much more than that to me. There are people I encounter on my journey through life who are more than friends. Our connection is strong, independent of time spent together. After not speaking for years at a time, we meet up and find we're just as close as ever. No matter how much either of us has grown or changed, we are still beautifully in sync. I call this "Joined at the Karma", and I believe Condon and I are privileged to share a bond that transcends time and space.

We also shared a number of meals, and after each he would pick up the check and expense it. You might think of this at putting the 'A' in the middle of ET, but we saw it as yet another enhancement to an already beautiful relationship.

Then there's the issue of my laziness. I'm so lazy, I only want to do tasks in the most efficient way. In fact, that is so appealing to me, I'm willing to do way more work than a task requires, just to wind up doing it in a better way. Perhaps you recall our earlier discussion of programmer brain. The satisfaction I get from doing tasks this way more than compensates for any "wasted" time or energy. This could mean I'm good at optimizing or bad at being lazy. Either way, that won't work here. Now I must choose one path and start running. I'm sprinting a marathon, there's no time for dilly-dawdling.

Will I try to work every waking moment of these 35.5 days? Unlikely.

Knowing me, some fun time will be required. All work and no play makes Jack Nicholson a snubbed Oscar nominee. I'll need to make room for at least one crazy binge, if only for sanity's sake.

TYPICAL DAYS

People ask me: "What was the hardest thing about doing E.T.?"

One answer is: Typical days.

Typically, there is no such thing as a typical day at Atari. Game makers experience a wide range of emotions on the job, from the exhilarating to the terrifying, but very little monotony or boredom.

Creating the E.T. video game will not be boring, but the day-to-day experience of production can be dreadfully monotonous. On this project there *are* typical days… and I hate that. Typical days are the second hardest thing about doing this game.

First place, however, goes to the reason there are no typical days at Atari: My colleagues and workplace. Be it fun side or dark side, this is my sustenance. Encountering remarkable people on a daily basis, finding solutions together, facing fears, sharing gossip. Doing a video game in five weeks is extremely difficult, but the hardest part isn't what I'm doing, it's what I'm missing. Cutting myself off from this amazing world I so love and need, *that's* the hardest thing.

Also, these particular typical days don't make for compelling imagery. Day in and day out, I'm just sitting, thinking and typing. Sure, inside my head it's very dynamic, but for spectators it leaves much to be desired. Of course, if my brain-speaker is still connected, you can hear the old Rawhide song playing over and over:

> Codin' codin' codin'
> Though the bugs are trollin'
> Keep those pixels glowin', On Time!

Sitting at a dev station day after day is not terribly dramatic. But the context, the implications, and a few moments along the way…

CHAPTER 11
SO MANY QUESTIONS

Any questions so far? I'm sharing what facts I have as well as my opinions/theories, but this long and winding road has plenty of detours and blind curves. I don't pretend to hold all the answers, but I do have lots of questions. Until now, I've studiously avoided dealing with the single biggest question of the entire E.T. video game project: Why only five weeks?

This one is easy, and one of the few times an Atari answer is straightforward. It was deemed the E.T. game had to be available for the Christmas sales market. Christmas can't be moved, so just start working backwards from there. In order to take proper advantage of a Christmas release, a game must be in stores by early November. Which means it has to be packaged and shipped by Halloween. Which means if you want to put four million of them out there, the finished game must be debugged, quality tested and delivered to manufacturing by September 1st. They didn't greenlight the game until negotiations concluded on the morning of July 27th (about an hour before I received the phone call). That's why there are only five weeks for the entire development. Makes sense, right?

[Joke for the Number-Nerd: Some people find this timeline confusing. They think Halloween and Christmas are the same day because Oct 31 = Dec 25. Admittedly, this is rather base humor. If you don't get it, you can google it. Google knows this joke.]

To understand this more deeply, we have to dive a little deeper into Spielberg's filmography. While the game itself was trying to live up to the E.T. movie, the schedule for the game was caught up in another Spielberg blockbuster… Jaws! With one key difference: That movie was about the jaws of a shark. My schedule was caught in the jaws of a vise. Your basic vise has two jaws: a static jaw and a sliding jaw. The static jaw is Christmas, that's not going to move. The sliding jaw, however, is the ongoing negotiations for the rights to do the game. In other words: There's no leeway in the finishing date, the only flexibility is when to start. Every day spent negotiating is a tightening of the vise jaws, squeezing precious hours out of the development schedule. There were only five weeks left to make the game because it took so long to close the deal.

Finally, having secured the rights, they immediately start calling engineering to see about doing the game. What's hard to understand (yet illustrative of issues that will ultimately kill Atari) is why management waited this late in the game to consult engineering? Why didn't they check-in *during* negotiations? Wouldn't it help them to have some idea about what it takes to make a game? Where was Jerome and his dioramas in this critical moment? Oh yeah, he was in my office reading National Lampoon letters to the editor. We didn't know we were giggling while the development schedule burned.

That's the real crime of the E.T. game. They spent precious development hours in protracted negotiations, time which could have been spent improving the game. E.T. didn't have to be the worst game of all time. It could have risen to the level of most mediocre game of all time! Another missed opportunity, quel dommage.

Doing a video game in five weeks is an incredible ordeal. If you ask an executive, "Why would you put anyone through this kind of strain?", the answer would likely be: "Huh?" They had no idea what they were asking, and I don't think they cared. They saw an opportunity and went for it. It's a game, how hard can it be to make? That's what development is for, you demand things and you get them.

Management believed they had the golden touch. In reality, they were just out of touch. Of course, with no awareness of resource or schedule requirements there's no reason to be concerned. Deliver a game in 5 weeks? No problem. We're Atari, we can do anything.

I can't help wondering, would negotiations have changed if they'd taken input back in June? Might they have nailed it down sooner? Adding just one week would have been a *20% increase* in the overall schedule. But that didn't happen.

Here's another possibility: What if the drive to make the Christmas market was born of something else entirely. Remember how Atari always has secret undercurrents and backdoor conversations you don't know about? Some execs might have an inkling things could be starting to go south in the industry. What if it isn't just wanting to make the Christmas market? What if we *need* to make this Christmas market because there may not be another? If so, it still seems a rather lackadaisical approach for a desperate act.

And given that management was willing to put me through this, why would I put myself through it? I guess it's hard for me to refuse such a monumental challenge. Add this to my as yet unquenched quest for validation and it becomes an irresistible force I simply cannot resist. So I didn't. After all, most great accomplishments (and outrageous calamities) begin with someone saying "yes".

Here's another reasonable question still hanging: Why me? Why *does* Howard get to do E.T.? Two reasons: First, Spielberg requested I do it because he was pleased with what I did for Raiders of the Lost Ark. Second, no one else in the world was willing to attempt it. I am literally the only person anywhere (including my boss's boss) who is willing to say I will deliver this product. That's why I was doing E.T. Oh, and one more reason: I happened to work at Atari at the time…

THE INEVITABILITY OF ATARI

You know that feeling of waking up and not knowing where you are or how you got there? It can be terrifying until you're reoriented.

Developing a video game for the VCS typically takes at least six months. I know this because I've done two games already. Yars' Revenge took seven months and Raiders of the Lost Ark took ten. Doing another one in five weeks? That may sound like a difficult programming problem, but it's not. It's an impossibility.

I contemplate this in conjunction with the fact I've committed to doing just that, and the terror strikes. How did I get here?

The truth is: My life was always heading toward becoming a video game programmer at Atari. This would have been obvious if I'd been able to read the signposts along the way, twists and turns which only make sense in retrospect. Fortunately, it's 2020 now, the Year of Perfect Hindsight! Looking back, it's easy to see how I blazed a trail leading inevitably to a place I barely knew existed.

In fact, it started before Atari existed. By the age of five I was sitting in front of a pile of broken toys and old game pieces saying, "I'm going to invent something. What can I make with this?" I never did, but the intention burned in me.

Do you remember class photos in elementary school? While handing out the pictures, my fourth-grade teacher liked to predict our future professions based on our portraits. When she came to mine, she looked it over and said "Engineer!" I stared and stared at that picture, but for the life of me I couldn't see anything that might make her think I would end up driving trains. It turned out she was prophetic. I'm embarrassed to admit this, but I didn't understand there were other kinds of engineers until my freshman year of college. It's a good thing I found out before grad school.

As you may recall, throughout my childhood and adolescence I loved games. I'd learn a game, play it for a while and then try to improve it. I found many games could become more fun or better balanced.

Arcade style video games started appearing just as I was entering college. I remember walking into a Blimpie's sandwich shop in New Orleans and seeing a Space Invaders for the first time. I looked it over and told myself: This is going to be really big! I didn't play it though. My friends got deeply into it, and I'd watch them play. They were so taken with it, but my focus was elsewhere.

Then one day, an interesting thing happened in my Calculus III class (now there's something you don't hear very often). I will spare you the gritty mathematical details, suffice it to say while hearing a lecture on complex surfaces, a full video game design came alive in my mind. It occurred to me I knew enough math to create a multi-player tank game. I envisioned players driving around a 3D landscape, hunting for (and hunted by) other players. It would be easy to calculate. Each player could have their own screen, and surprisingly little information would need to be shared, so the communication could happen very quickly. It would be an efficient "real time" interactive multi-player networked computer tank game. This idea is commonplace today, but it didn't exist in 1976. And this was before I had any exposure to computer science whatsoever. I honestly don't know why I was thinking this, but I was. Video games just made sense to me, even if they didn't captivate me, yet.

Clearly, Atari was a place I was heading. But in order to arrive there I'd have to get involved with computers, and I nearly didn't...

As an undergraduate at Tulane I had a double major in Economics and Mathematics with a minor in Theater. I was going to be an economist. Computers were not on my radar. In fact, I avoided them because that was nerd territory. Don't get me wrong, I was a total nerd (except for the stereotypical social shyness), but I was not yet ready to self-identify as a nerd. One fateful day I was chatting enjoyably with my Economics adviser. She told me: "You won't get anywhere without computers." OK, perhaps it's time to give computers a try. It's the middle of first semester in my sophomore year, and I don't want to wait to start this new adventure. So, I start thinking about how I might hack the system and get into a class right now.

It turns out I had done so well in my freshman year that I was designated a "Tulane Scholar". So far, the only benefit I'd found to being a Tulane Scholar was getting invited to a wine & cheese at the president's house during the first week of school. When I inquire further, I am told Tulane scholars are entitled to certain academic advantages and expediencies, none of which were ever named. I take it upon myself to innovate. I find out who is teaching the Intro to Programming course and pay them a visit.

Dr. Victor J. Law is my cultural antithesis. He is all about the south, and his neck is as red as a neck can be. He loves nothing more than an afternoon BBQ or department party where he can sit on the ground and drink wine until he loses the power of coherent speech. But he is much more than that. He is also brilliant. Dr. Law earned his Doctorate in Chemical Engineering by writing an algorithm so complex yet so elegant that no one knew how it worked. But they all knew it worked and constituted an important contribution to the field. However, Chemical Engineering is not where Dr.

Law's head is at these days, because he is also ambitious. He is interested in spearheading a brand-new Computer Science department in the school of Engineering. In the late 1970's, this was the thing to do in universities, and Dr. Law is always on the lookout for good candidates who might populate his new endeavor.

Enter Howard, a high pressure, fast talking, New York accented, dyed-in-the-wool yankee invader. I walk right up to his desk and introduce myself as a Tulane Scholar, which entitles me to add his course in mid-semester. Well, he looks me straight in the eye and tells me exactly where I can go... to find the textbook and the computer lab. Years later he will share his delight at giving this [expletive deleted] [epithet deleted] just enough rope to hang himself.

But I don't hang myself. I get the book and go to the computer lab and finish the first seven weeks of the course that night. A few days later I have finished the course. OMG, it's a revelation! Computer Science is my academic dream come true. I get to solve puzzles and never have to write papers or read long meandering books. I absolutely eat it up. The more I get, the more I want. Everything about it feels like a perfect fit. Even the giant fanfold computer listing paper has a remarkable fringe benefit: It's perfect for removing the seeds from marijuana. (Does anyone remember when pot had seeds?) Though I'll finish my Math and Econ majors, it's clear they will no longer pave the road to my future. It's a programmer's life for me. Better yet, it looks like there's way more job opportunities in computers than in economics.

I've finally found my direction, and Dr. Law has found someone who will constitute 50% of the first graduating class of the Tulane University Graduate School of Computer Engineering (the other half being a delightful New Orleans native named Archie Greffer, a very cool guy who makes The Big Easy a little easier). The yankee and the redneck have found each other, forming an unlikely but mutually beneficial partnership. We each understand the value of our relationship. He sees in me an opportunity to get his department off to a good start, and I see in him an opportunity to upgrade my bachelor's degree to a master's before venturing out into the working world. And it comes to pass that our wishes are granted. His new department is launched, and I am conferred upon in a Master-ful way.

My enhanced degree catches the eye of Hewlett-Packard in Cupertino, California. They fly me out for an all-day interview and make me an offer before the sun goes down. I accept on the spot and my future is assured. At long last, real life beckons.

When I report for work at Hewlett-Packard, I am very much an engineer. I have a master's degree in Computer Engineering which I received at the age of twenty-one. I'd have done it sooner, but I didn't discover computers until halfway through my bachelor's degree. That's how I approach things, I avoid them as long as possible and when I can't avoid them any longer, I dive in. My high school had a terminal connected to a computer at Rutgers University. I could have started working on computers in 10th grade. Instead I avoided them like the plague. It wasn't until late 1976 that I first touched a computer and found out I liked it. By May of 1979 I had a master's degree. If something is worth doing, it's worth doing fast. Especially if I waste a lot of time before realizing it's worth doing.

It is also worth noting that my specialty in grad school was microprocessor-based real-time control programming. This was very rare back then, but more significantly it is the work I've come to love.

So, what am I doing at Hewlett-Packard? Having a crisis. The raging passion which led me to this grand life circumstance has dissipated. They don't do microprocessor programming. They do bigger, slower, less engaging projects on mainframe (big) computers. Over time, I'm realizing that HP doesn't feel like a hotbed of excitement and innovation. To me, it feels more like a software pasture where programmers go to finish out their coding days before they die. I'd gotten what I wished for, but it wasn't what I needed.

Yet again I find myself unaware of where to go but clear it isn't here. Though familiar, this is not a good place for me. I'm deeply saddened and disappointed, so I start looking for opportunities to find joy in my day, or as others might put it: Acting out.

We have color-coded clips for marking interesting parts of large reams of computer listings. I string them together into sizable chains and use them to decorate my desk. There's a networked star trek game which many people in my group play after hours, I get very involved with that. The football pool provides some distraction in fall, playing racquetball and poker with coworkers is good too. I'm pouring myself into everything except my actual job responsibilities. It's hard to face the sudden absence of compu-joy. It's hard to face myself wasting valuable time as I am. Passion is my fuel, but I can't catch even a whiff of a fume. This can't go on much longer. It turns out this is one more way life is guiding me toward Atari and video games. I just can't see it yet.

I'm not a nose-to-the-grindstone type. I can fake it for a bit, but giving my all requires genuine inspiration. That's not happening at Hewlett-Packard. I'm entering the first of three great depressions. It's a very difficult time for me, but sometimes life kicks you out of a comfortable bed so you don't miss an appointment with destiny.

I'm sitting in my review with my manager, Bill. He is a wonderful person who sees tremendous potential in me. It's hard to watch him struggling to get it out of me. It's hopeless. I *feel* hopeless. All my work and effort in college led me to this miserable moment. I am totally empty. For the first time since childhood, I'm crying. Tears of desperation fill my eyes and stream down my cheeks. I've reached the pinnacle of "supposed to be here" and it holds nothing for me. I cannot continue this charade. I need to go, and I have no idea where.

Within days a ray of light breaks through in the form of Vince, one of my cubicle colleagues. He told his wife some "Howard stories" and they remind her of where she works. Oh, where's that? Atari.

In the darkest hour, my roiling frustration sent ripples out into the Universe like radar signals, bouncing around for a while and eventually returning to deposit a singular blip on my screen: Atari!

Roger that, Universe.

I reach out and they bring me in for many interviews. It seems to go well. I give them the Howard show and go through their whole process. It turns out Atari does microprocessor-based real-time control programming, and they use it to make games. More than a good job fit, this is my specialty, my passion and my long-awaited answer. The moment I fully grasp the situation: Instant nerdgasm!

With a background like mine, it's easy to see why I would be a game engineer for Atari. What's harder to see is why they rejected me. It's true. After this entire process, Dennis calls me to say they won't be offering me a job. Really? I don't think so...

My entire life has been moving inexorably toward this moment, but not so fast. Of course there's another hurdle. I must defeat the boss before passing through the final portal. I'm certainly not giving up now. I'm too far through the looking glass and the White Rabbit is anxiously tapping his pocket watch. Must. Work. There.

I ask Dennis why the rejection? He says they think I'm too strait-laced for their environment. This will become a very funny joke in the years to come, but no one is laughing now. I understand immediately what is happening. This is the ironic part of the reversal.

Atari is an extremely unconventional environment, and I'm an extremely unconventional guy... normally. But I goofed. During the interviews I dressed nicely, tucked in my attitude and adopted a professional demeanor. You know, like one does on interviews. Big mistake! I misrepresented myself as too conventional and they misread me, leaving Dennis with the impression I'd be a fish out of water, lost and misplaced at Atari. Sadly, that's exactly how I am right now in the software pasture. At Atari, I'd be a fish very much IN water.

For me, this rejection is just a kickoff to negotiations. I get the miscommunication and I recognize my contribution. I start by sharing my fervent belief we may be missing an incredible job match. He acknowledges this possibility and the game's afoot. He cites problem after problem, and I counter with solutions at every turn. Eventually I wear him down and he agrees to give me a shot, provided I accept a probationary period and a 20% cut in pay. I shout: "Deal!" And with that, my fait is finally accompli.

[NOTE to the Triviologist: "The Game's Afoot" is also the title of a book written by Ken Ludwig. It was published on November 14, 2012. This is the exact day I became a licensed psychotherapist in California. Now the head-game's afoot!]

[NOTE to the Sequentialist: The previous note may seem a tad non-sequitur, but the truth is it came out of nowhere. Truthier still, it was the product of a spontaneous Google search. Ah Google, is there any task you can't make longer by stopping to smell the trivia?]

IN, YET NEARLY OUT

Though my arrival at Atari seemed inevitable, there were two times I came cluelessly close to not being around to pick up the phone when E.T. came calling.

I'm not talking about potential firings, I'm referring to episodes in which, wisely or not, I attempted to leave Atari. Fortunately, some of my best life decisions were made for me, not by me.

You may remember from tales of Christmi past, that many Atari programmers were made merry by generous boni in late '81. This is a parting gift from young Rob and Bob and others, since it seemed to proceed from their leaving to form Imagic months before. Though I hadn't been there long, I am saddened by their leaving and a little hurt I'm not among them. These are some of my favorite people.

Two months later, Brad Stewart (my officemate and the other half of the Atari pun-itentiary) also leaves to join Imagic. My sadness deepens a bit. I can't help feeling if only I'd been here a little longer, they might have wanted to take me too. I also couldn't ignore the fact Imagic offered them a much more lucrative deal for making games.

I finish Yars' Revenge and begin working on Raiders of the Lost Ark. As exciting as it is to be working on my first Spielberg property, the grass seems so much greener on the Imagic side of the fence. I even go so far as taking an interview with them. In an extremely secret meeting, I sit down with William F.X. Grubb, another former Atarian who is now co-founder and CEO of Imagic.

He explains to me how I would be a valued addition, but they cannot consider bringing me on. He and all their ex-Atari people are already spending too much time in legal proceedings as it is. Hiring me will antagonize Atari further and may expose them to yet another lawsuit for poaching programmers.

Interestingly, this is another marker of the cultural shift at Atari. Under Nolan, Atari tried to accomplish its goals through engineering. In Ray's Atari, the legal department gets a good deal more exercise. And in this case, those muscles are getting the job done.

I don't feel the need to push back here like I did a year earlier when Atari refused me. Though I do enjoy a challenge, there is always something to be said for choosing one's battles. I decide to accept the situation and settle back into working at Atari, which currently consists of working with Steven Spielberg on the first ever movie-based video game. Those grapes are not exactly sour.

Given the opportunity, I would have gladly gone to Imagic. And if you recall the last of the three Christmi, this would have found me on the shareholder side of the epic undercutting of their IPO. One more Imagician with coal in their stocking.

Instead, life continues as if that super-secret meeting never happened. Nothing changes. I stay put and play Atari lotto. Will my initial disappointment harden into regret? Or might this rejection be a blessing in disguise? Honestly, to this day it's still not clear to me.

And then there was the second time I almost left before the E.T. project came up…

THE DAY THE WORLD CHANGED

Thursday, February 25th, 1982 was a day that changed everything. As amazing Atari days go, it's one of the all-timers. It started out plainly enough, at least by Atari standards. I wake up excited, knowing I have a plan of my own cooking on the burner. A plan to change my life in a major way. A plan to which I'm totally committed. A plan which, by late afternoon, will entirely disappear from my mind.

As I arrive at work, I see the all-hands notice. Apparently, there's a spontaneous mandatory all-programmer meeting this afternoon. It's likely about whichever recent violation of decorum received the latest complaint. Few things make a day feel longer than bureaucratic gestures disguised as technical meetings. Whatever. I'm mostly focused on my own plan. Plus, there is plenty of work to do on my current project, Raiders of the Lost Ark.

Atari doesn't know it yet, but as of two days ago Raiders of the Lost Ark was facing a serious development issue: The potential resignation of the assigned programmer. I may be leaving Atari sooner rather than later. As of this moment that is my plan, and the reason is money.

Money had increasingly become an issue at Atari, which is ironic because none of us came here for the money. I certainly didn't. I took a 20% cut in pay to come to Atari, and I did it gladly. It was all about the joy. The challenge, the atmosphere and the fun, those were the draws, and they drew a most interesting crowd. We didn't come for the money, but there was no denying these games were generating a tremendous amount of money and very little of it was going to the people who made the games. It's also a relatively rare thing that a product with such high profit potential is produced so inexpensively by individuals. Product quality and sales figures made it relatively easy to sort out the talent. Whenever people left to form other companies, it was usually the most talented people and it was mostly about the money.

So, it is no surprise that my plan, the one about which I've been obsessing for the last 42 hours or so, is also largely about money. It's also about Tod. After the Imagic formation, Tod and I formed a strategic partnership of our own; we agreed if either of us was approached by venture capital, we'd include the other. Well… earlier this week, Tod was approached by 20th Century Fox who was looking to be the latest entry into the explosively profitable home video game market. He came to me and we agreed to do this together. We also agreed not to discuss it with anyone else, as this was highly sensitive information in an already prickly atmosphere.

I've been kicking this around in my head for a couple of days and right now I've got Rob-Fulop-scale visions dancing around in my head. No mandatory meeting is likely to interrupt my reverie.

I'm sitting in my office, timesharing between Raiders code and visions of what my new office in the new company might look like, when Tod walks in and shuts the door. For some reason I'm reminded of my first day at

Atari, another time Tod walked in and shut the door. His offering today is significantly less fragrant.

Tod plops down in the swivel chair, seamlessly transitioning into a spin. On his second full rotation, he shares with me the conversation he had yesterday with our grandboss George Kiss, head of our department and Director of VCS development. It seems he told George about our plan. Smiling, he recounts their exchange: "You know, George, I really like this job, but it's beginning to cost me money." Yet another classic Tod line. The man has a flair for drama.

Though I appreciate his delivery, I do not appreciate the content. We agreed not to say anything that might mess this up. At Atari, phrases like "I'm leaving to form a competitor" are litigatin' words. But Tod is going to do Tod, and any expectation to the contrary is either denial, ignorance or foolishness… a trio I prefer to avoid. The fact is, if you want to step into Tod's world, you must do so on *his* terms. But there's no denying this conversation makes the day feel even longer still, as well as a bit more paranoid.

Fortunately, there's little time to wallow in the mire as the time for our mandatory meeting has arrived. Tod gives the chair a final whirl and off we go to the meeting.

As we enter the room, I see Mike Moone. Mike is president of the Consumer Electronics Division, in which we are all cogs. Also present is Ray Kassar, our CEO. These are the biggest of wigs, and their presence is somewhat warping the concept of "all-programmer meeting". Something big is going on. I think back to Tod's revelation moments ago and a sinking feeling comes over me. This day just keeps getting longer and longer. Is this really where I want to be right now? I check the door for security guards, seeing none there I consider running. Suddenly I'm seized once again by the realization of how remarkable it is that I'm even here right now.

You, dear reader, are the only other person in this room who knows I went over to the dark side and interviewed with Imagic two months ago. And as you also know, they rejected me. That hurt, but I pretended to understand.

My current plan is young and hasn't matured to the "I'm gone" phase, at least not yet. Has Tod's recent maneuver slammed that door shut as well?

Damn you, Tod! If only Imagic would have taken me with them. If only Tod had gotten this going earlier. If only I'd interviewed at Activision (an idea I never considered until just now). But alas, here I am, watching the band assemble, waiting to face the music.

Of course, this isn't exclusively the fault of others. After all, Dennis did his best to reject me, but I wouldn't have it. I could not take no for an answer, because I needed to wind up… here? With so many exits on the highway, how did I miss them all? I am nearly despondent as the meeting comes to order.

It starts out with Ray Kassar walking around the table and handing envelopes to people who had released games within the last couple of months. I qualified with Yars' Revenge apparently. Was this some sort of expression of gratitude and appreciation? I must admit, I half expected to open it up and find a certificate for a free turkey.

A moment ago I was dreading what may come of this moment. But one thing I have learned is not to pass judgement on current circumstances until I see where they take me. Sometimes life looks out for us despite our worst impulses. I'm here now because my attempts to not be here have failed, so let's see where I am…

I open the envelope and inside is a check. A check in the amount of $28,120.00 which I'm being told is $40,000 minus withholding. Ray goes on to explain that there's a new Atari Bonus Plan in town, and it pays some real money. Here's the proof. It has completion bonuses (of $40K) and royalties on our games! For the first time it seems like Atari is coming through with real value. Finally, the people who make things will start making money from the things they make.

The first thing I do is examine the check as carefully and thoroughly as I can. $40,000 in 1982 is a LOT of money. That's $109,350.64 in 2020 dollars. Put another way, If I'd invested that money in the Dow Jones index at the average price in 1982, that 40 grand would now be worth over one million dollars. Spoiler alert: I didn't do that.

I verify the check is made out to me, and that my name is spelled correctly. I think about checking my driver's license to ensure the names match. Ray and Mike go on explaining the plan, and I'm very interested in hearing about this plan. Later I'll pore over the document in excruciating detail, but for now… Mind. Blown. I'm way too busy in my head right now to pay any attention to what's being said.

I'm holding the biggest check I've ever seen in my life and I'm positive it's made out to me. This new plan, well, that's the only one I'm interested in now. I glance over and catch Tod's eye. He's holding a check too and it looks like we are of the same mind. Neither of us cares about 20th Century Fox anymore… unless it's movie night.

Next it occurs to me that this is exactly the reaction Ray and company are looking for, and I'm cool with that. But I also remember this is Ray Kassar, Atari CEO, the man who will never pay towel designers or high-strung prima donnas as a matter of principle. This is truly an odd happenstance, bordering on twilight zone.

I begin to realize why today feels so long… it's another one of those extra-long days. It turns out February 25th, 1982 actually started on October 1st, 1979 to be exact. That's the day Activision, Inc. was founded by engineers who wanted more profit participation in their products. This of course smoothed the path two years later for the Imagic formation, where the next round of Atari engineers left to increase their earning potential. And now it is about eight months later and Ray just got wind that Tod and I are preparing for round three. Here's what I think really happened…

The long-awaited inception of this real Atari Bonus Plan was as simple as 1-2-3. First there was Activision which merely pissed Ray off. Second came Imagic, another drain of his stable. Since neither Activision nor Imagic had faltered (despite Ray's most aggressive legal attacks) he's starting to get worried. This leaves one more shoe to fall, and it drops when Todd tells George about our plan. The alarm bells go off and the fire brigade springs into action. There is screaming, swearing, flailing and ultimately some check cutting. All this leads up to this meeting. I believe that Tod's pronouncement merely reminded Ray of that classic management maxim:

Fool me once shame on you.
Fool me twice shame on me.
Fool me three times and I don't want to have
to explain it to Warner's Board of Directors.

With outside developers, Ray's philosophy was: If you can't beat 'em, buy 'em. It just took a while to accept this applies in-house as well, otherwise you can wind up with more of an out-house experience.

I took a moment to bathe in the irony. Despite all my best efforts to leave, those failures left me here to receive this check and a plan which would make my Raiders of the Lost Ark project very worthwhile indeed. Sometimes when I stumble and fall, I wind up dodging a bullet. I was lucky this time, and all because Tod broke our agreement and spilled the beans. Thank you, Tod!

Now I feel I'm working with a clearer picture of the whole situation, especially the part where I realize this is not bad for me at all. In fact, it's rather fantastic! As the relief pours over me, I settle back into the room where the Ray and Mike show is just wrapping up. Having taken leave of the meeting for so long, I realize I'm way behind my smartass quota for the meeting, but I find my stride quickly enough.

I look up to the head of the conference table, hold my check up in the air and say, "You know what I'm going to do with this, Ray. I think I'm going to start my own company." The braying ass strikes again. I don't mean it, of course. On some level he may know that too, but clearly it irks him.

The therapist in me will come to recognize how this reflected the resentment engineers held for management due to management's refusal to acknowledge the importance of engineering. As in all relationships, resentment is just a symptom of the dysfunction that's already there. My healthier therapist-self would not make this comment. It's always easier to be mature in retrospect.

In this moment, however, I am here now. I did say it, and I'm considerably less healthy than my future self will be. In my defense, I'd like to say there is really no defense for this. It's a dick move on my part for sure, but

OMG, you should've seen his face.

This changed everything in the environment, and not just for the programmers. Marilyn, the first Art Director in video game history, really knew how to represent for her peeps. She was responsible for getting royalties for the digital artists/animators in her group. Remember in that department meeting when everyone was so upset because they felt I got all the big titles? Now you know why. It wasn't just pride, it was price as well.

Tod and I never go with the 20th Century Fox venture, which is fortunate. They end up finding some other programmers who do go with them. And guess what? Fox Games folds shortly thereafter, and those programmers end up coming to Atari for jobs. And there, but for the grace of Tod, would've gone I.

CHAPTER 12
LUMP IN THE SNAKE

Back in the desert, the weather has eased up for now. People are mulling about and chatting enjoyably. Production hands scramble around. Lines form at the food trucks as people grab a quick bite during the lull in the storm. News crews rearrange equipment while the on-camera correspondents practice their interview lead-ins and set-ups, all in preparation for capturing whatever may unfold in the next few hours. Thankfully, the airborne sand and debris have relented. The still air is now filled with a palpable anticipation. We are waiting to see what ancient secrets this desert repository is willing to reveal.

Legend has it Atari not only dumped millions of unsold E.T. games here, but also ran bulldozers back and forth to crush the cartridges into uselessness and then topped it all off with a thick layer of cement. That's a lot of care and attention (and money) to sink into something no one wants. Could Atari bury millions of my game cartridges in this dump and keep it under wraps? Not possible. I fancied myself attuned to the company grapevine. I'd certainly have heard about it. I'm sure several people would have loved to give me this news. Of course, the entire negotiations for the game were done quietly enough that no one in engineering knew about it. Perhaps my Atari vineyard connection wasn't as fancy as I think.

However, if anyone *had* told me, I know exactly what I'd do: Hire a photographer and fly us both to this site. An 8x10 glossy of me perched atop the desert grave holding millions of my games? That's a must have! It would go on the wall in my office, right next to my Romper Room diploma.

You know, I could take that picture right now. It would be presumptuous since I don't yet know if the games are here or not. The weather above ground has cleared up, but the ambiguity surrounding what lies beneath the surface remains.

[NOTE to the Romper-Room Nerd: Yes, I am a certified Good Do-Bee, appearing on the Romper Room show at the ripe old age of five. My big start in show business. Perhaps if I'd listed this on my resume, Atari wouldn't have rejected me in the first place. Despite holding multiple degrees and certificates, my Romper Room diploma (complete with photos of my appearance) was the only one I displayed at Atari. Thematically it

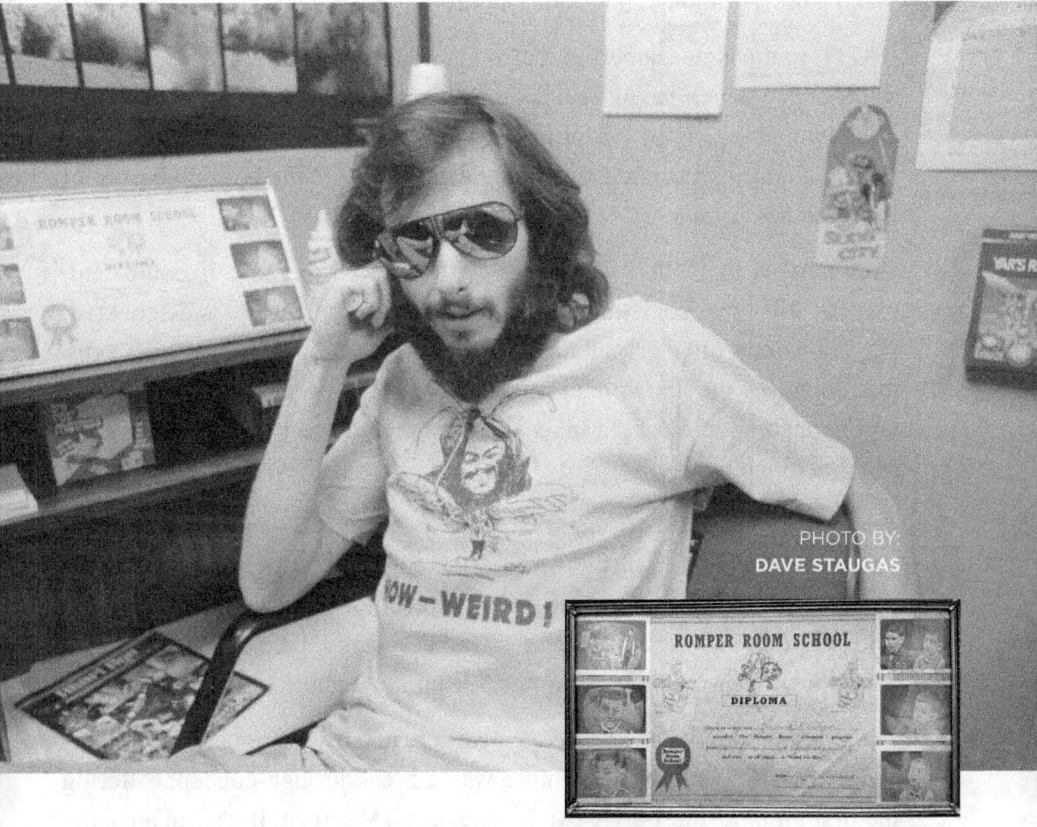

was just the right touch, and it was practical too. Many good doobies were rolled on this Good Do-Bee certificate, and many lines of cocaine disappeared from its surface (much to Miss Barbara's chagrin, I'm sure). Recreational pharmaceuticals are clearly a Don't-Bee activity.]

GAME HUNTING

This dig has a lot going on. There is a lot of fun to be had, a lot of pressure to get results, a lot of adventure to enjoy and a lot of money at stake… it's just like Atari.

Standing here in the warm desert air of Spring, I remember last Sunday was Easter. This reminds me of another kind of hunt, perhaps we should be searching for Easter eggs… also just like Atari. Placing Easter eggs in video games is a long and proud tradition.

[NOTE to the Non-Gamer: A Video Game Easter Egg is an object, image or sequence which has little or nothing to do with the basic game action. It is obtainable only by performing some obscure action which is unlikely to happen during normal gameplay. Easter Eggs are the buried treasure of video games.]

How cool is that? Very cool. Now, after the basic game is completed, players still have something extra to seek. And once one is discovered, players wonder: Is there another? It's a lagniappe, a little something extra that enhances the appeal and mystique of the experience. Like a flower you happen upon during a hike; It's not why you came, but it makes the trip that much better.

When you find an Easter egg in a game, you can trust it was put there for your entertainment. But the first Easter eggs, the ones hatched at Atari by the dawning industry's early light, were not there for entertainment. They were there because of trust, or rather a lack thereof. Easter eggs are magnificent flowers that sprang forth from the manure of distrust.

The original video game Easter egg was created by Warren Robinett in his VCS game, Adventure. Adventure was a breakthrough concept, offering the first graphic quest-style experience on a TV screen. It was an amazing feat in just 2K of code, and it created an entirely new genre for gamers. That was his contribution above the surface.

Hidden below the surface is a secret code sequence. When executed correctly, the player is rewarded with a strip of text running down the screen: "Created by Warren Robinett." So, why do it?

True or not, in Ray Kassar's Atari, the suspicion someone might be out to screw me over was never far away. They even hired Nixon's former head of security. Talk about paranoia.

In the beginning, Easter eggs were secret signatures designed to show authorship. Activision and Imagic highlighted their game developers by putting their names right on the box. Atari wouldn't do that. Atari games are made by Atari, period! You may recall Yars' Revenge was the first Atari game to have a programmer credit (albeit in the comic book). It would never happen again. I was able to break the rule, but I couldn't change the policy.

Atari developers were supposed to remain anonymous, which was a sore point. Here's why: Let's say I tell someone (a friend or prospective employer), "I made this game for Atari." They say, "Really? That sounds like bullshit." I say, "I did! Trust me." But what if they don't? How can I prove it? With an Easter egg, I can say, "Hand me the controller…" and I can make something happen that nobody else can. Ideally, it's something no one else would put in the game, like my name. That's what Warren did in Adventure, and it was brilliant.

Not everyone did it, but many did. And in a variety of creative ways. From obscure play sequences to odd switch configurations. One programmer told me they put the computer codes for the letters of their name in the game memory. You must extract the program and view it with a code editor to see it, but it's there. Most people were more playful with it. Take me, for instance.

From the moment this piece of lore was handed down to me, I was all about the Easter eggs. It's like Tod Frye says, "We're engineers. What we do is develop a system and explore its capabilities... and then we develop new features." I loved it, and simply could not resist becoming an Easter egg-gineer. I not only wanted to sign my games, I wanted them to cross reference each other.

Yars' Revenge only has one signature. If you're in the right place at the right time, you will see an HSWWSH appear instead of your score. This is significant to the game in that it's the key to understanding the naming convention behind the game. These are my initials forwards and backwards. This tells the player to spell other things in game backwards, like Yar and Razak (Ray Kassar).

Raiders of the Lost Ark has two signatures. You can find an HSW2 at one point, which indicates it's my second game. And you can also find a Yar flying up the screen in another location. I decided each of my games should contain the main character of all previous games.

On E.T., the player can find a Yar, an Indiana Jones and an HSW3. Also, E.T. may be the only VCS game which also has a signature for the graphics/animation designer. There is a way to make a JMD appear in the status indicator. These are the initials of my good friend Jerome Domurat. Additionally, if you look carefully at the three phone pieces in the game, you will notice each one is a deformed letter. There's an H, an S and a W.

OK, I don't know if you're thinking it, but I am: Gee Howard, for a game that needed every second it could get, you sure spent a lot of time and effort elaborating extraneous egocentric eccentricities. Wouldn't that time have been better spent planning, permuting & proffering productive product possibilities?

Your point is well taken. I beg your indulgence while I construct a rationalization for my indulgence…

The cost of doing it? Not much. Although it sounds like a lot of different things, they were all the same activity: healing the flower under the right circumstances. As for the phone pieces, that was Jerome's time. All I did was ask him to make the phone pieces that way. My policy on signatures: They don't go in until late in the game. They don't get much memory and they must be quick to program. All together, these took less than two hours total.

The value of doing it? I knew there wasn't enough game play in E.T., so I figured having more Easter eggs might increase the life of the game (now *that* is a rationalization).

The real value comes from the fact that I enjoy putting in signatures. It's one of my favorite things to do on a game. I save it until late in the project when I'm draggin' and it perks me up. Doing this game in five weeks was grueling. It wore me down and I needed something to bolster my spirits. Adding a signature always gives me a lift. It's like in the movie The Producers (the 1967 one), when Zero Mostel takes time out from his exhausting old-lady seduction marathon. He opens the safe and starts sniffing the cash, thus renewing his motivation and enabling him to continue his arduous journey. But I digress.

Another thing about early Easter eggs: Putting a signature in your game is strictly against company policy. Therefore, the prime directive when implanting an Easter egg is keeping it secret... at least until you get through manufacturing.

When it was discovered that Warren had Easter egged his Adventure game (a player found it and wrote a letter), Atari management was quite miffed. They don't seem to recognize the added value Easter eggs contribute. Of course, there is nothing they can do about it because no one outside engineering can read code, so on it goes.

I put a secret signature in Yars' Revenge, hidden behind a play sequence so obscure that no one would ever find it accidentally, or so I believed. I was obeying the prime directive, but it all seemed kind of silly to me. Why are we hiding game improvements from the very people who are looking to profit from them? I decided to break another rule and try to get marketing involved in the process. There's really no downside to it. If they reject my proposal, I'll just leave the secret signature in anyway.

I go talk with Jewel, the gem of marketing. I lay out the whole story about signatures and Easter eggs, then I propose they should be celebrated additions to our games rather than a source of internal conflict. Essentially, let's make chicken salad out of chicken shit, and possibly enhance our game appeal in the process.

Marketing approves the proposal. I change the super-secret sequence into a much more accessible configuration, and we work it into the manual as "The Ghost of Yars."

I broke the rules, but that's what innovators do. Now Easter eggs don't have to be secrets anymore. There were so many secrets at Atari. I always found it interesting that a place with so many secrets had so little trust. Don't secrets depend on trust?

To be clear, I'm not saying there was no trust. I trusted Jewel, she came through and it worked out great. I trusted Tod, he broke that trust and that worked out even better. But there were others at Atari who inspired more dubious feelings in me, especially in the context of deal making. Like the day the black limo showed up...

REACHING THE RIGHT AGREEMENT

It's nearly the midway point in the E.T. project and I'm at the office. I come in two or three times a week because in 1982, sneaker-net is the most reliable upload/download technology. I share my latest version of the code, get a fresh listing and synch up with Jerome to see how the graphics are coming along. Things feel tight but under control. So far, so good.

[NOTE to the Non-Nerd: Sneaker-net is a data transmission technology in which I place the data on some storage media (i.e. disc, thumb drive, tape, etc.) and physically walk it over to another location. It was very popular before the internet existed. A Listing is a paper print-out of the code in a program. They were usually printed on huge fanfold paper sheets whose utility ran far beyond programming.]

I'm chatting enjoyably in the animators' room when the receptionist comes in with a quizzical look on her face and says, "Howard, there's a car here for you!?"

THE ANIMATORS' ROOM. LOTS OF FUN AND CREATIVITY HAPPENED HERE!

I glance at my watch, notice it is the appointed time and tell Jerome I'll be back later. Jerome thinks nothing of it, but he wouldn't. He's pretty unusual. I go back to my office, pick up my latest listing and head to the front lobby. There I'm met by a gawking gauntlet of curious faces. Word travels fast around these parts. There's a shiny black limousine waiting for Howard and everyone wants to know what's up? Fortunately, they're asking with their eyes and not their mouths, so I figure I'll just show rather than tell. I walk through the onlookers, out the door and get into the waiting car.

A quick check-in with the driver confirms he's clear on our destination. As he pulls down the road toward the onramp to Highway 101 north, I settle back in the plush leather seat for my 45-minute excursion. After watching a few trees go by, I pretend to start examining my listing. I say "pretend" because my brain is mostly focused on that initial phone conversation with Ray Kassar. Technically, just one small part of that exchange, the one that precipitated this limo ride:

"Howard, we need an E.T. game for September 1st. Can you do it?"

"Absolutely I can! Provided we reach the right agreement."

I must work virtually every waking moment if I'm going to deliver the game, but putting this off is unacceptable. I haven't yet caught my breath from the exhaustive ten-month slog through Raiders of the Lost Ark, all done while navigating the steep highs and lows of suspicion, exhilaration and Atari intrigue. Jumping immediately into this ordeal is taking a toll on my health, both physically and mentally. Is this too much? Will I become another casualty? I need to ensure doing this to myself will be worth the effort… and the risk.

I'm right in the middle of the project. My leverage is large now, but just like a python's last meal, it shrinks every day as it moves through the schedule. Once they get the goods, my leverage is gone. I can't let that happen. It's time to have my own secret negotiation.

Well, semi-secret. I'm heading to a lawyer's office in San Francisco. Perhaps I could have found a lawyer somewhere closer, but this one comes highly recommended.

I'm taking a consultation in preparation for my upcoming meeting with Atari executives and lawyers to "reach the right agreement." I believe the most important part of any negotiation is being clear about desires and expectations. So, what exactly do I want?

One thing is money. I must admit, there was a time in late June where I had been soured by a discussion with an executive who shall remain nameless. They tried to sign me to a deal which sounded like it would increase my workload and reduce my pay, then criticized me for not seeing how much this would benefit me. That pissed me off. I told them something to the effect of: I'm not quite that big of an idiot. Then I looked them straight in the eyes and said, "I know you'll want me to do E.T. next, and that's gonna cost you!"

Did I really know I'd be doing E.T.? Not at all. I pulled that out of my ass in an attempt to needle them. Besides, I'd just seen the movie. In the vernacular of the time (as well as an ironic foreshadow): I pity the fool who thinks E.T. should be a video game. Raiders of the Lost Ark? Now there's a movie destined to be a video game. But E.T.? Forget about it. I spent zero thought on any E.T. game design, but that exec had irritated me enough that I did start contemplating a deal.

In my fit of pique, I picture myself sitting across the table from this exec and demanding $2 million dollars, one for me and one for taxes. They say yes and I win. It's all good. This is entirely fantasy of course, I don't know of any effort to get the license. Interestingly, the exec may know of it at this point, but I don't. And even if they had the rights, it's already too late to hit the Christmas market.

The limo rolls up the peninsula toward the city by the bay.

The one where Tony Bennett's heart remains this very day.

How absent minded could he be to leave his organ there?

As I approach this self-same place, I'm suddenly aware… that I'm sitting in the back of a hired car, mindlessly thumbing through my fanfold listing. Slowly my thoughts drift back to the present, entirely sans pique. So, now that I'm living the unforeseeable, how much money do I really want?

This is a potentially precedent setting moment. I'd heard rumors of game engineers past who sought special deals for doing games and none of them worked at Atari anymore. But I'm in a unique position no game engineer as ever been before. Two million sounds nice, but it doesn't sound realistic or appropriate to me. I'm not trying to gouge anyone. There's no guarantee this will work. The only guarantee is extreme duress and strain on my part. I need to ensure a reliable yet reasonable upside for attempting this craziness. But what does reasonable mean in an absurd world?

Since the new bonus plan came out, we are guaranteed royalties for our games. If this game gets released, it will likely do well enough to generate a good chunk of money. Perhaps the best approach is an advance against those royalties? A non-refundable advance. It becomes inconsequential if I deliver but still assures a nice bonus for the effort if it all goes sideways somehow. That sounds reasonable to me. Right now, I likely have more leverage than any VCS programmer ever has, but I don't want to make any enemies. I just want to make sure I don't pile drive myself into the ground for nothing in the case of a near miss. But how much should it be? When I think of setting precedent, my mind usually turns back to the same place: Yars' Revenge.

That game knocked several bricks out of the of Atari thou-shalt-not wall. The programmer's name is in the packaging, it displays some of its own code on the screen, and that pesky Easter egg is not only in, it's blessed by marketing. Shouldn't Yars assist in removing the next brick from that wall? And speaking of bricks, it occurs to me that players receive 169 points for eating a brick from the shield in Yars' Revenge. How about $169,000? If it's good enough for Yar, it's good enough for Ray.

[NOTE to the Atari Terminology Nerd: Though it is true the Yars' Revenge manual refers to the shield elements as "cells," I always thought of them as bricks. I certainly had a lot of input to the manual, but I did not write it.]

OK, that's fine for the money. What else should I seek in this agreement. How about security? At least some Atari-style approximation of security.

This latest iteration of the ever-evolving Atari bonus plan is quite generous, it's active, and I am mightily happy with it. So why am I insecure? Well, there is that "ever-evolving" thing. You never know when the next bonus plan is coming along, and what if it constitutes a retreat from current levels of remuneration? That can happen too. Atari giveth and Atari taketh away. I've been here long enough to recognize a double-edged sword when I see one. Another clause worth adding to the agreement is one to preserve my current bonus plan in case Atari changes it down the road. A little CYA.

That feels better.

I believe I've got the basics of a reasonable deal in my head. This begs the question: If I'm so clear on what I want, why bother with a lawyer? Because I'd like to ensure the agreement we draw up represents my ideas faithfully, without any slippery loopholes. I'm not trying to hold them up. I just want to nail down a deal I can trust, and trust can be a hard thing to come by at Atari.

We arrive at the building and pull into the underground garage. I locate the elevator and begin my ascent. After gaining altitude for quite some time, the doors open to reveal an opulent reception area. It seems their offices occupy the entire floor. The receptionist leads me to a conference room that blows my mind. Two of the walls are floor-to-ceiling windows, offering a stunning panoramic view of San Francisco bay, including the Golden Gate bridge, Alcatraz, the Transamerica pyramid and the Bay bridge. It's so clear I can see Berkeley. Light blue sky. Rich green foliage. Dark bay water with bright white breakers. Breathtaking!

After too short a time, someone else collects me. We walk down a long corridor to the next venue, which will also take my breath away.

I enter a huge office where five people are already positioned and waiting. Three are middle-aged serious-looking suits, each mastering the art of appearing authoritative yet remaining deferential. Then there's the young associate, far younger than the others. I imagine he's the fast-rising phenom, top of his class, now catapulting himself into San Francisco's legal elite. I imagine this because, had I gone to law school instead of engineering, that's who I'd be trying to be (I believe I made the right choice though). Last but not least, posed impressively at the focal point of the room, is the white-haired patriarch of the firm. He's leaning against his throne and puffing on his pipe. The blooms of smoke are so imposing, most of the oxygen in the room has already given up and left. Introductions are made, hands are shaken, and I'm offered a seat right next to the young associate. I feel he's been selected especially for me, like a translator for a foreign dignitary. "Make sure that new guy Clarence is in the meeting. He's hip, or whatever they call it these days, right?"

Now the meeting proper begins. The patriarch doesn't really talk so much as orate. He gets things rolling, issuing abstract dictums on negotiation and points of law. They sound great but I fear they contribute little, other than adding expensive tics and tocs to the time meter. Between clouds he underscores his points with elaborate arm gestures, each leaving in its wake a trail of un-puffed smoke from the glowing pipe in tow. He goes on a while, billowing and bellowing. Then, as if on cue, the youngest associate leans in and starts talking to me. It seems he is tasked with turning these broad abstractions into nuts and bolts content for my benefit. As he's talking, it strikes me I'm sitting in the lawyer scene from one of my all-time favorite movies: Network. The dialogue today isn't nearly as sharp, but it's eerily similar. [Mental note: Next time, bring my own writers.]

It doesn't take long to dispense with the nuts and bolts and proceed directly to the screws...

The associate fires the next salvo, "Let's ask for one million dollars."

"That seems like too much. I don't want to antagonize anybody."

"But you're holding all the cards right now."

"Yes, but I want to preserve a good working relationship."

"This is an incredible opportunity. You've got to push while you have the advantage. That's why you have us."

"The deal I have is fine. The reason I have you is to ensure we sign an agreement that will let me keep it."

His face is oddly contorted by confusion and disbelief. He cannot comprehend the concept I don't want them to nail Atari to the wall. They want to push the envelope and I'm restraining them. This exchange leaves me with an important question: Am I an idiot?

I'm paying someone to hold back and not fight harder for my benefit. In my mind, I'm just trying to secure what I already have. After all, I'm making lots of money doing something I truly love to do. I'm living the dream, yet I'm feeling the need to defend myself from the very people who are making this possible. What's wrong with this picture?

As my therapist, I would suggest this could be a chance to work out my resentment over management's undervaluation of product development. In other words, they are so self-centered, they don't understand how important I am! This is my chance to get their attention in a way they can't ignore. I can hit them where it hurts, right in the money. But I'm also dealing with my own undervaluation of me. Against my lawyers' advice, I don't really put the screws to them. I believe the parade is going to go on for a while and I'm just trying to preserve my place in line. I'm very idealistic about my situation and the future.

Unfortunately, I'm also very wrong. Though I didn't think so at the time, I was very naive. In retrospect, maybe I should have gone for a bigger cash grab. Alas, hindsight is so easy in 2020.

The actual meeting with Atari's lawyers and execs went off without a hitch. Well, one hitch. They want everything in the agreement to be contingent on a delivery date. They say if it's not delivered on September 1st, then the deal is void. I get it. They want some consequence if I fail. But it raises another question, what does delivered mean? If I'm nowhere close, then that's fine. But what if the game is basically ready on September 1st but they don't accept it until the 2nd? I wouldn't put it past them to hold up approval until the day after delivery just to nullify the agreement. Wow.

An air of distrust pervades everything management presents to engineering. It may sound odd or ridiculous, such unprecedented cynicism in young and well-paid employees. But we saw Activision and Imagic form and we knew why they did. It's an interesting phenomenon.

Anyway, their ask is reasonable, I just need to adjust it a little. I stand firm with the non-refundable advance against royalties, but I'm willing to grant the rest of their request provided we change the condition for delivery. I propose Spielberg as the ultimate arbiter. He comes up on September 1st and if he says the game is done then it's done. I figure, if anyone is motivated to ship this game ASAP, it's Spielberg. They agree and the deal is done.

The idea that my proposal was met with immediate assent is a sure sign I asked for too little. A wise man once told me: If they accept your first offer, you came in too low. But I came in right where I wanted to be. The wisdom is debatable, but my heart was pure and my conscience was clear.

I also believed there was a lot more runway left before the cliff. Had I known the reality, I might have done it differently. On the other hand, I clearly remember thinking: If Atari is my only contribution of note, I probably don't deserve the big score yet. I have more to do.

Over the years, I've had occasion to reconsider my demands. More money? That might have been nice, but money comes and goes, I could easily have messed that up. There is one idea that keeps creeping into my mind though, and it's the only thing I truly wish I would have added to the deal:

Rather than more money, I should have asked for more time. Not time to do the game, but time with Steven. Getting one afternoon a year with Steven Spielberg, every year, as long as we're both around. That would have been a great contract clause as well as a great gift (at least for me). And as long as Spielberg has to sign, I'd also ask that when he (and his secret scouting crew) take off, they take me with them. Don't worry, Bob, I'll explain that reference after I finish the game.

Since time is of the essence, my lawyer writes up the agreement. We go back and forth a couple of times, and by week's end the deal is signed and I'm holding a check for $169,000.

I made the special deal in the dark. Another milestone achieved. Atari lore is haunted by ghosts of those who had tried and failed. You can't always get what you want, and no one is indispensable. But in the entertainment (and techno-tainment) business, exceptional value commands exceptional leeway. Standards, rules and limits are for average contributors. If you want others to suffer your eccentricities, overlook your transgressions and generally tolerate your bullshit, just be profitable enough that they can't afford to lose you. Then you can do whatever you want, as long as you deliver! That's the secret formula. No one is indispensable? In the long run that's true. But when dependencies are huge, reputations are on the line and corners are painted into… maybe not so much.

For now, seeing all those zeroes on this slip of paper in my hand, it's a good day. I believe it's time to cherish the moment and celebrate!

CHAPTER 13
THE HOME STRETCH

Three weeks down and two to go. I've hit that classic point in software projects: I'm 80% done with 80% to go. It's a great place to take a break, something I haven't done since the project began. Usually this is an opportunity to relax and reflect on the journey so far. A chance to reconsider the best path forward. But backing off and reevaluating is not on the menu today.

The clock is ticking. Time is short. I need a break to avoid crashing and burning before I reach the finish line, but it has to be a quick one. In this predicament, the best I can hope for is blowing off some steam as efficiently as possible. The operative word here being *blow*.

SNOWING IN AUGUST

I need a mini vacation. Some short term high-velocity reliable excitement from which I may return seamlessly and resume the crunch. It's binge time. I put the word out to fellow coworkers who constitute the mainstay of my microcosmic world. Arrangements are made, supplies procured, back-ups assured, and the group gathers for a white out.

Events like this would happen on occasion. They weren't regular or frequent by any measure, but they happened. These gatherings were a relatively recent phenomenon too. I believe the first one I attended came shortly after February 25th, the day the world changed.

As money flowed into VCS engineering, certain activities got upgraded (at least from our perspective). Bocce lemons graduated to remote-control car demolition derby. Modest economy vehicles turned into high-performance sports cars. And at party time, marijuana could morph into cocaine, as was the fashion in the early '80s.

We'd get together, lay out lines and talk the night away. Sometimes all night and even into the morning if the mood strikes and supplies hold up. The great thing about coke is the illusion. Under its auspices, every thought is a fabulously brilliant breakthrough, and that is a nice feeling. It's the most fun you can have sitting in one place for ten to fifteen hours in a row. But it *is* an illusion.

The reality is: I simply wind up poorer and wanting more. Some find the white powder useful for fueling extended work sessions. This was never my style. I always kept my programming separate from my pharmacology. For one thing, I don't trust myself to do complex programming tasks when I'm wasted. I'm afraid I'll create more problems than solutions. More importantly, I don't want my productivity linked to a substance. My use was strictly recreational, and knowing my own proclivities, I found it wiser to binge and then quit for a good while. Engaging this particular chemical on a regular basis felt far too dangerous for me. Luckily, I never lost that feeling.

So, for one magical evening I put everything aside and hosted an escape hatch. A mini marathon of sparkling conversation, flashes of blinding insight and nasal spray. Eventually I was able to follow that up with a good sleep and a nice meal. After my 24-hour vacation, I was ready to hop back on the horse and continue the ride.

THE FINAL WEEKS

Atari gives me the chance to do everything that inspires me: Facing and meeting big challenges. Overcoming obstacles and hurdles. Coming up with new ideas and fresh ways of doing things. Having impact and creating positive change. Helping people in a real and meaningful way (like healing or entertaining) and being well paid for it.

Creating this E.T. video game is the greatest inspirational opportunity of my life so far. It's the chance to realize an impossible dream, and I'm doing it. I'm almost there…

With the exception of my mini-vacation and the surreal encounters during negotiations, I'm working all the time. At home, at Atari, even while driving between home and Atari. You can bet your life I'm working on the game while I'm driving, because I certainly did.

I edit code in my head as I head down the road. It turns out that can be distracting. While sitting at a stop light waiting for green, I unlocked a particularly tricky sequence and now I can't wait to get home to type it in. The light turns green and I hit the accelerator. As I bolt into the intersection, I hear brakes lock up and tires screeching. The light is green, but only the left-turn signal changed. I'm going straight. Fortunately for me, the oncoming left-turner is not coding in their head. They note my miscalculation and stop in time, which is good because I don't. With no police witness present I make it home just as the adrenaline is clearing my system. I run to my dev station and type in the changes before I forget them. Only then do I admonish myself to be more present and focused while driving. After all, careless lapses like that could delay the project unacceptably.

That was the second most dangerous moment of the project. The first is a far more profound example of programmer-brain and the risk factors involved when intense focus meets tunnel vision…

Despite the aforementioned admonition, less than a week later I am once again cruising down Lawrence Expressway on autopilot. My brain is coding away like a good little time bomb, when something rather odd comes into my awareness. I must admit, when I'm driving I have a habit of ignoring things like signs and lights. Instead I focus on the things that can hurt me, specifically my fellow moving objects. I'd rather run a red light because there's no cross traffic than cruise through a green light and get hit by an oncoming vehicle, but that's me. My brain reads vehicles first, signs and lights second.

As I'm drive-coding down the road, my programming reverie gets interrupted by something curious. I'm approaching a major intersection and I can't help noticing the cross traffic. The cars nearing the intersection don't appear to be slowing down. They're just rolling along at consistent (and considerable) speed. This strikes me as rather odd behavior for people coming up to a red light. I glance up at the light, looking solely at the light in their direction since that's the one that matters for the moment. Well, what do you know? It's green, so of course they aren't slowing down. That explains that.

Only then does it occur to me: Since they have a green light, mine is probably red. Another quick glance confirms my suspicions. I mash the brake pedal into the floorboards. My brakes lock. My tires screech. I'm skidding as the car is fishtailing. After an eternity of fear and dread, I come to rest on the far side of the (thankfully unoccupied) crosswalk. Holy shit!

There was no one ahead of me approaching the intersection or I would have reacted to their brake lights. I've always believed the most pertinent limit is the speed of the car in front of me. And there is no one directly behind me since they were all quicker to note the redness of our light and behave accordingly. It's just me, the unwitting stunt driver dangerously on display. I shrink down in my seat, melting under the stares and glares of other drivers coming to rest nearby. I'm waiting for the telltale siren blast foreshadowing impending doom, but none follows. The only sound is my heart mercilessly pounding my eardrums. Sadly, this is a familiar beat lately. Sadder still is the realization that the adrenaline from these road events is the only exercise I'm getting during this entire project.

The light changes. No one moves. Everyone is waiting to see what the unpredictable maniac will do first. I hit the gas and get the hell out of there, taking an alternate route home. Though I don't know it, I'm incredibly grateful that cellphones aren't *a thing* yet. Absolutely no programming takes place in my head for the rest of the ride home.

The road to "extreme" is paved with limits, and the potholes are getting nastier as I'm exceeding mine. Too much tension, too little sleep. Too much adrenaline, too little rest. It's a volatile mixture. This project lit a fire under me, but after four weeks on the rotisserie I'm thoroughly crispy… and "charred" is fast approaching.

With one week left, the game is now playable. It's raw, but playable. Testers are working it over and early feedback is starting to flow. There is concern over some of the game mechanics, but the messengers are getting nervous because I'm starting to snap at people, which really isn't typical of me.

The pressure is overtaking me. I'm running on fumes, and I'm resenting it. How dare my limitations intrude at this critical juncture!

Yes, it's easy to stumble into the wells. I understand it feels annoying at first but that's *gameplay*. The challenge is mastering tight maneuvers to skillfully guide E.T. through narrow margins. This is the morsel of gameplay in my design, I can't afford to abandon the sole test of eye-hand coordination. Can't they see that?

I'm trapped in a limited vision, resisting any and all change. The transaction in my head goes like this:

> *It's not a fatal bug that crashes the game. It will take time and thought to address this issue and there's no room for either. Racing to the finish line is all that matters now, and I can see it. Playability must be an afterthought.*

I'm ashamed to admit I have unwittingly adopted marketing-mindset, and I'm totally on board with it. I am going to deliver a completed… something. And it's not like this is my last game, right?

The simple truth is: There's too much to do and too little time. A game can always branch off in a hundred different directions, each with its costs and benefits. I'm driving a high mountain road with a sheer cliff. In every moment, the most important thing is to keep making the choices that keep me on that road and not falling into the chasm. Deviating from the plan right now risks everything. Crispier still.

Thinking back on the moment I answered Ray's phone call and chose to launch this entire fiasco, I realize the only two words I heard in that conversation were "Learjet" and "Spielberg." I was hooked. How could I not be? I'm working at a job where the CEO calls me up and says: "We need you to get on a Learjet and fly to Hollywood so you can talk to Steven Spielberg." *That's* my job. The day I sauntered into Dr. Law's office and requested enough rope to hang myself, this is where I was heading.

Moments like this make it crystal clear I'm in the right job in the right place at the right time *right now*. As insane as all this is, life is perfect, and I LOVE IT! Suddenly all my stress melts away and I feel renewed... for about three minutes.

This snapping at feedbackers is a symptom of my burn out and abject frustration. It's no secret this game suffers from an appalling lack of time, but the real cost of this micro-schedule is tuning and rumination time, critical factors in any solid development. Tuning is about refining what's there and rumination is about new possibilities.

Game changing innovations come from mental background processing and the odd stray thought. Serendipity is a valuable asset in game development, but it can only help when given the opportunity. Having more time allows one of nature's fundamental forces to come to our aid: entropy!

[NOTE to the Non-Nerd: Entropy is a principle in Physics which states that every system tends toward randomness unless energy is spent to impose order. Entropy guarantees if I don't clean my home it gets messy. If I don't organize my desk it becomes an ever-deepening pile, held in place by an infinite number of pending sticky notes. Though it's also true that if I don't eat I get hungry, that isn't entropy. That's just dieting. Dieting is not physics. We know this because physics actually works.]

[NOTE to the Physics Nerd: Don't give me any shit about that definition of entropy. It is a perfectly functional lay-definition, and it is clearly labelled for non-nerds. Respect the specifiers! And why are you reading other people's notes, eh?]

[NOTE to the Note Reader: My good friend Bob would like you to know that a more colorful explanation of the value of tuning and rumination time is coming up later in the book. And it features a special guest appearance by René Descartes.]

OK, back to serendipity and entropy. I enjoy the irony of needing randomness to improve my design structure, but I'm not smiling right now. The price of a five-week development is the near-total loss of tuning and rumination time. Of course, this is not news, it was absolutely clear from the start. At least it was clear to anyone whose hubris was not casting them into total denial.

"I'm really good at this. I can do a game in five weeks. I'll figure it out." The BMOBS emitters in my brain have been redlining from the start. Remember BMOBS? I believed my own bullshit. The problem now is that reality is finally piercing my denial, poking its ugly mug through the veil just to laugh in my face. Reality can be such a dick at times. After going two-for-two on hit games, I'm hitting a wall on this one. I remember when Yars was in trouble. That felt bad but I still had months to address the issues. Now I have just days to finish and my programming dance card is already full.

Clearly the chances of delivering my usual quality of product are reduced to an extremely low order of probability. This is acutely stressing me and threatening my sense of competence and self-worth. These are manifesting as irritability and anger, the secondary response to the primary emotions at work: fear and pain. At least that's how I will come to look at it after decades of distance, reflection and a degree in clinical psychology. But right now, each time I hear something is subpar or problematic with the game, all I can hear in my head is, "Oh shit, I'm screwed. Keep paddling, it will be better when it's done. Oh shit, I'm screwed. Keep paddling, it will be better when it's done. Oh shit, I'm screwed. Keep paddling…"

There's an old joke that goes like this: A software engineer, a hardware engineer and a salesperson are on their way to a conference when they get a flat tire. They pull over and discuss what to do. The salesperson says, "Let's buy a new car." The hardware engineer says, "Let's rotate the tires and find out which one is flat." But the software engineer says, "Let's just keep driving, maybe it will fix itself."

Old jokes get to be old because they tell the truth. In this moment I am living the joke, but I'm still not smiling. I could just tell myself, "Gee Howard, it's not working out after all. Oh well." But that's not where I go. Instead, my thoroughly stressed, sleep-deprived and paranoid brain goes here:

Time is short. Yes, there are issues, but the plan is adequate. I just need to finish. It's doable if I stay on course. Any fundamental change now will cost a week or more of time I do not have. Stay on target. Stay on target. Stay on target… That makes sense, right?

To paraphrase Leonardo da Vinci, Creative projects are never finished, they are simply abandoned. On this game, my bags are packed and I'm planning to run away at the first opportunity.

I ignore all input that is not a total crash or significant bug. Frustrating gameplay is relegated to that sacred programming maxim: "It's not a bug, it's a feature." The headwinds blow, but I press onward.

Finally, in the late evening of Aug 31st, disheveled, depleted and limping, I drag the E.T. video game though a successful testing cycle and across the finish line.

I DID IT!

Of course, this is merely *my opinion* for now. Tomorrow we will all find out if it is, in fact, a fact.

WEDNESDAY, SEPTEMBER 1ST, 1982

After a thirty-five-day gestation and a week of intense labor pains, the game is crowning and delivery is at hand. In today's performance, the role of the midwife will be played by Steven Spielberg. He is here to pass judgment on whether the baby, my baby, is healthy or not. I'm playing the nervous dad. I didn't sleep too much last night. I'm conditionally looking forward to seeing Steven again. He will evaluate (and hopefully accept) the E.T. video game, clearing it for the hand off to manufacturing. All the tension, turmoil and craziness of the last five weeks comes down to this moment.

Little do I realize this will be the last time I'll be together with Steven Spielberg for a very long time. This seems like a good place to tell you about the first time we met…

It's late summer of 1981 and another amazing day is starting. I get up way too early (of course), grab a Yars' Revenge cartridge and hustle to the airport (no Learjet this time). The mission: Fly down to Warner Studios for a 9:30am meeting with Spielberg. This is basically an interview for the position of doing the first Movie-to-Video-Game conversion in history. The movie to convert: Raiders of the Lost Ark. Great movie! And quite a challenge on the VCS.

I catch a commercial flight to Burbank, then a taxicab to the studio. As the cab works its way through the LA morning traffic, I'm beaming. This is so cool! I remember watching Duel for the first time and thinking "Wow, what a great way to tell a story." I remember spilling my popcorn when the face floats into view in Jaws. Close Encounters filled me with wonder and dreams and visions and ideas of all sorts. It grabbed and inspired me. I got goosebumps when those five tones played. Steven Spielberg doesn't just make movies; He creates significant moments in my life. I want to do that for people too. He's one of my heroes. I feel the tingle of goosebumps washing over me once again as it sinks in that I'm on my way to meet him.

The guard at the gate actually has my name on his list! He gives me directions and a smile. I make my way to Spielberg's office by 9:25am. I pause to gather myself after this harried morning journey. Composed, I take a deep breath and open the door, right on schedule. Here's the receptionist in the anteroom, the gatekeeper.

"Hi, I'm Howard Scott Warshaw and I'm here to meet with…"

"Hello Mr. Warshaw. Your meeting has been rescheduled to 3:30 this afternoon."

"Seriously?"

"Yes."

"But I flew here from San Jose for this meeting."

"Yes, and I know Mr. Spielberg is looking forward to talking with you."

Honestly, I'm a bit miffed. It turns out I didn't need to get up at oh-dark-thirty and scramble to the flight so I could catch the cab enabling me to arrive just in time to SIT AROUND FOR 6 HOURS!

But my irritation doesn't last too long. I'm more of a make-lemonade kind of guy. Most people consider me an optimist, and that's a fair assessment. By the way, do you know the difference between an optimist and a pessimist? The Optimist believes this is the best of all possible worlds, the Pessimist fears this is true. At least that's what Robert Oppenheimer said, and he was the bomb!

I'm definitely optimistic, but I'm really more of an opportunist. I'm thinking: How can I *make* this the best of all possible worlds?

First thing I need to do is change my flight, and I'm standing right in front of an assistant. Hmmm.

"Can you change my flight for me?"

"Certainly. Just give me your tickets."

Oooh, that was fun. But now I've got six hours to kill. What's next? It occurs to me I'm standing in the middle of Warner Brothers studios, and I happen to be a huge movie and television fan...

"Is it OK if I just walk around the studio until then?"

"Sure, go ahead."

Really? I get to spend a day roaming (unescorted) around a major film/TV studio?? Then I come back and chat enjoyably with Steven Spielberg??? This is truly the best of all possible worlds. After being genuinely miffed just moments ago, I'm already "over myself" and setting off on a magical adventure.

Lemonade, bitches!

I start wandering around the studio grounds doing anything and everything I feel like. I go in and out of sound stages. I roam the back lots, passing facades of old New York streets and a western town. I hang on the periphery of several outdoor shoots for a bit. The only thing I don't do is enter any door with a flashing red light above it.

As lunchtime nears, I notice hordes of oddly paired types moving in the same direction. I flow with them and wind up at the commissary where I eat with aliens, sheriffs, clowns and knights in shining armor. After lunch I continue meandering about. Studio tour trams roll by and I hear tourists speculate about me...

"Is he anybody? Do you recognize him?"

"Nah. He's probably just an extra, or an exec."

"I could be an extra exec," I think to myself, but I say nothing. I decide the role of GUY ROAMING LOT is not a speaking part.

I come upon a huge building with an unguarded door. The red light above is unlit. Dare I? After glancing both ways I test the latch. It's unlocked. The door is now slightly ajar, it's practically an invitation.

Inside is one of the biggest insides I've ever seen. I wander around. No one else is here. The space is partitioned. Cables and equipment are strewn about here and there. I come around a dividing curtain to find I'm on a set. The back wall is painted in a tropical way, there are lights and wind machines around, and there's a flowered trellis... Boss! Boss! I'm on Fantasy Island! This is one of my favorite shows and now I'm *on* it, kind of. How perfect is this? I'm having a fantasy day and I wind up in Fantasy Island.

As I look closer, the trellis barely stands on its own and the flowers adorning it are the cheapest of plastic. But they look so beautiful on TV, I guess I'm through the looking glass now. I pretend Spielberg and I are guest starring in an episode. Art imitates life. In short order I decide it's time to move on. But before I go, I steal one of the flowers off the trellis. I hope the statute of limitations has run by now.

Not once in six hours does anyone stop or question me. I guess show biz is in my veins. I'm. Loving. This.

As the appointed hour draws near, I conclude the amblin' part of my adventure and return to Spielberg's office. As I approach, my heart beats a little faster with each step. Is this it? Is it Spielberg time? I open the door and the helpful assistant informs me that nothing is rescheduled. She ushers me in for the main event and the next thing I know Steven Spielberg is shaking my hand. He looks just like him!

We spend some time chatting enjoyably. After all, we are both suburban kids with the need to create. We find a quick and easy rapport. Perhaps because he has the eye of a child, unspoiled and deeply in touch with the wonder of the world. I believe maturity is not the negation of childhood but rather the integration of it. I suspect he agrees. Our inner children get along famously.

I don't like to fawn, but I do share how much Close Encounters means to me. He mentions how he's starting production on another movie featuring a friendly character from outer space. That's interesting.

We talk for a bit. We play Yars together. It's going fine and he seems to like my work. Then I decide it's time to share a thought.

Bear in mind this is the early '80s, and it really feels like we're getting close to encountering our galactic cohabitants. Bear further that I happen to be sitting with one of the people primarily responsible for this aura of possibility, which is incredibly cool for me. I say…

"Steven, I have a whole theory about how you are actually an alien yourself. Would you like to hear it?"

He smiles. "Sure. Let's hear it."

I lay it out for him: "It really feels like we're getting close to the first alien contact. But I think when the aliens arrive, they won't just show up in a spaceship and say: 'Here we are!' I think it's more likely we discover they're already among us. If they're smart enough to find and approach us, they're probably smart enough to do a little recon and prep first. So, I figure they'll have an advance team whose job it is to prepare earthlings for a smoother transition. This team will have two components; production and marketing. The production arm creates media more favorable to the alien image than we've traditionally seen on Earth. The marketing group then ensures these new messages are seen all over the globe. After movies like Close Encounters and this current project, I figure you are the production arm. And obviously the marketing group is doing some amazing work, since Close Encounters was seen all over the world in every language. So, I just want to take this opportunity to say, nice job!"

He doesn't give me a lot of feedback in the moment, but he's not kicking me out of the office. I take it as a good sign.

We talk and play and brainstorm for a while, then I take a cab back to the airport in time to catch my rescheduled flight. I'm floating from a perfect day. I got to hang around a major studio, eat at the commissary, get mistaken for a movie exec and to top it all off, I not only get to meet my hero, I get to connect with him. I retrieve my Fantasy Island souvenir from its secret hiding place deep in my pocket. Holding it just below my nose, I sample its delicate aroma. No plastic flower ever smelled so sweet.

Upon returning to Atari the next day, I'm informed my next project will indeed be Raiders of the Lost Ark. I believe my alien theory is a major reason why. Months later, Games magazine will call me up to get the scoop (Spielberg mentioned it to them during an interview). I relay the story and end up getting quote of the month for calling Steven Spielberg an alien. To this day, it still tickles me.

That was my first meeting with Steven Spielberg. Today, just under a year later, we're here for another close encounter. Once again I'm presenting a finished game for his approval, and this time there's a lot more riding on it.

It's a big day at Atari, you can tell by the crowd. There's the entire VCS management team who frequent our halls, and some midlevel execs who seldom our halls. Ray Kassar and Steven Spielberg are present, as well as Jerome and me of course. Pleasantries are exchanged all around and in short order we file into the delivery room to see what's coming out.

The cartridge is plugged in and reset. E.T.'s enormous face is gracing the screen as the compu-tonal version of the John Williams score plays. I launch the game and run a demo for Steven. After a while I hand the controller to him and away he goes. As tense as this moment is, I still enjoy watching him play. His body English seems to suggest he is thoroughly engaged, and not a single swear's worth of frustration. It's early yet, but so far, so good.

I peruse the rest of the onlookers. Ray Kassar couldn't appear to be less interested, though I believe he cares greatly about the outcome. He is set back from the action, whispering away with a fellow suit. Perhaps he's just confident it will go well. Condon, George and the other VCS managers are far more focused on the main attraction, but no clear indication of their take at this point. The door jamb is jammed with half-faces peering into the fishbowl. Everyone is curious. How's it going? I catch Jerome's eye. With a slight elevation of one eyebrow and the nuance of a nod toward the screen, I request his take on the proceedings so far. Nonverbally he confirms he has no idea at this point. It hasn't crashed, Spielberg's still playing and the controller has yet to be thrown. So far, so good.

I don't know if it's lack of sleep, the relief/tension of sitting on the verge of verdict or just my usual nonlinear tendencies, but suddenly my head yanks the wheel into an abrupt left turn…

OMG! It isn't like it used to be. Where is Alligator-clip Jim!? What I am witnessing here, playing out before my very eyes, is a supreme example of Atari cultural transition. Allow me to explain.

One of the biggest questions in video game development is: How do we know when a game is done? This question raises more questions. Is it when all the promised features are implemented? Is it when consumer testing scores high enough? Is it when the design team feels it's right? When quality assurance signs off?

There is no single way to do it. These are all valid criteria and every game publisher sets their own process for designating a game complete and ready for distribution. This process is usually a reflection of the corporate culture. Atari is no different. We have release criteria too. Today, the release process is to bring in a major Hollywood film director and if he says it's done then it goes. That's not the standard company policy, but a symptom of the current policy. The point is: Atari's release policy isn't what it used to be.

The original release policy under Nolan Bushnell's leadership was: When the release committee says it's a fun game and does not suck, then we release it. The "release committee" is simply the game engineers. Those who make the product (and consume the product) decide when the game is done. That seems sensible.

When Ray Kassar took over, the criteria changed. Slowly at first, but over time the shift gained momentum. The key difference was transferring authority for release decisions from development (quality focus) to marketing (schedule focus). Don't get me wrong, of course marketing wants quality games and development wants games done expediently. But when push comes to shove there's a big difference between release-it-now and improve-it-first. And as licensing increasingly takes center stage, schedules just scream louder.

I do not yet know what this game may mean or become, but it stands now as a monument to the cultural transition from Nolan to Ray. The ultimate icon of the schedule-over-quality mindset.

And here we are. Steven Spielberg, who is only peripherally involved with Atari (or video games in general for that matter) is now the sole arbiter of whether or not this game is ready for public consumption. And I'm the one who insisted on it. My bad.

I was so blinded by my own short-sighted paranoia, I forgot entirely how little they care about the actual game. It never occurred to me that their attitude (which I deeply resent) works entirely in my favor here. To me, the saddest part is how I completely missed the irony. Ray never cared about my Spielberg caveat for release, I'm the smallest of potatoes to them. Nothing is going to stop this train from leaving the station on time. Well, one thing might…

As I merge back onto the cognitive highway, I see Spielberg is still engaged with the screen, but now he's doing more commenting than playing. I take this to mean he is reaching his fill and the verdict is likely forthcoming. He stops, thinks for a moment, then he turns to me and says: "It looks good. Let's go with it."

My inner world just sits there.

It's a remarkable moment. Steven is shaking my hand. I experience the exhilaration of acceptance and completion, while simultaneously falling into my own pit. Five weeks of adrenaline spontaneously drains from my system. I'd pop a cork in celebration but I'm too depleted to lift a bottle.

I'm absolutely empty, and in light of my recent driving record, I decide someone else will chauffeur me home today.

But I'm not leaving just yet. There's one more thing to do before a Spielberg game delivery is complete, and Jerome will help me do it.

There's a picture of me with Steven Spielberg in which we are shaking hands as I'm handing him a personalized copy of the Raiders of the Lost Ark game. Jerome created the graphics for "Steven", and I switched them with the copyright logo. This way, when he starts the game he sees his name on the screen. It is time to do the same thing for E.T. It won't take very long, but I can't feel complete until a personalized E.T. game cartridge is nestled safely in Steven's hand.

STEVEN SPIELBERG & HSW
PHOTO BY: DAVE STAUGAS

[NOTE to the Sceptic Who Has Actually Played the E.T. Game: On the off chance you are wondering: Yes, Spielberg did fall into some pits while he played the game. And yes, he did have a little trouble getting out at first. But he got a handle on it in short order and moved forward. Truth be told, it was rather impressive.]

Less than an hour later I place the freshly created "Steven" special in the palm of Spielberg's hand. There is no photo this time, but I do take a mental snapshot that has yet to fade. And with that, Ray escorts him out the door and the VCS department returns to abnormal… with one key difference: Spielberg's pronouncement has turned my opinion into fact. The impossible is accomplished and the die is cast.

On one hand, this game is a triumph of guile, determination, and the human spirit. On the other hand, it is a historic reenactment of the classic Ayn Rand quote: "You can ignore reality, but you cannot ignore the consequences of ignoring reality."

But for now, the consequences are nowhere in sight.

The game is done. Long live the game!

For the second time this summer, I'm giving Steven Spielberg a personalized game cartridge. This one contains an alien he made famous. I have no idea this same alien will make me infamous.

CHAPTER 14

KUDOS, BURNOUT & LETTING GO

The game is done. Now it begins!

Having pulled it off, I am a BMOBS star. I may also be a traitor to my cause. Accomplishing this feat may validate management's worst instincts and make it OK to continue this kind of scheduling, only time will tell. But when it comes to digging a deeper hole, I volunteered to be the shovel. It's poetic that three decades later I will wind up at a dig to recover the game, but I don't know that yet.

I haven't the slightest idea I've just delivered the worst game of all time. All I know right now is: I'm done. I did it and I need a break! For the first time in my Atari life, I need a real break. If only I could allow myself to take one…

THE AFTERMATH

I finished the handoff on Wednesday, then spent yesterday at home sleeping, eating and watching TV. Today is Friday and I'm ready to head back in and bask in the doneness. I have earned a monster birthday present for my 25th year: A victory lap. I arrive at the crack of noon, just in time for lunch. I spend the afternoon bathing in the afterglow of release.

It's been quite a while since the company threw a Friday party. A staple of the Bushnell regime, they pretty much disappeared about a year ago. Another casualty of the cultural transition.

This Friday is special though, it will host a spectacle of another sort: A company-wide meeting to celebrate the completion of the E.T. game project. This was a first in my Atari experience.

I arrive to find hundreds of people gathered in a particularly large room, some people I know and many I don't. Ray Kassar and Mike Moone are at the head of the crowd. I know them. Mike calls for attention above the din and things come to order. They say a few things about how daunting this challenge was, but we made it happen. And what a big coup this is for us. Then Mike singles me out and calls me up to the front. I join them and Mike explains how I came through for the company by pulling off this herculean feat. Everyone applauds. Mike reaches into his jacket, pulls out an envelope, and says: "As a special bonus, we are sending Howard to Hawaii. Here are two plane tickets to Maui!" I'm smiling wide now. The last time these two guys handed me an envelope it made me incredibly happy! I accept the envelope, but don't open it because that would be crass in this situation. I just hold it up like a trophy and express sincere gratitude at the top of my lungs.

It's a beautiful moment and people are cheering. Let's take a brief timeout and go behind the curtain... Both Ray and Mike know about the deal I've negotiated for the game. Since they know I don't need this extra accommodation, it occurs to me this is some sort of demonstration to show everyone else how they take care of people who come through. It's an interesting mix. I feel used, but in a delightful way. And I love the idea of Hawaii.

Shortly thereafter things wrap up and the crowd disperses. I take a moment to open the envelope and check out my tickets. But there are no tickets, just a small sheet from an Atari memo pad. It reads: "See my assistant for details. -Mike" Suddenly I'm transported through time and space, finding myself in New Orleans on May 12, 1979…

It's my fourth and final springtime in New Orleans. Today is impossibly muggy. Well, not impossible, since nearly every day it pours around 1pm and by 2pm the ground is dry because all the water has already made the state transition from liquid to humidity. After an hour in an air-conditioned class, the moisture condenses on my skin the moment I step outside. That's springtime in New Orleans.

Ordinarily I'm walk-racing to my next appointment so it's not too bad. But I'm stuck in a chair in the sun right now, draped in black from head to toe. My cap & gown may be slimming, but right now they're stifling. The graduation speeches go on forever. I guess I'll simmer 'till summer in all this pomp and circumstance. They should do this at 9am, when it's cool. But right now, I don't care.

I'm too excited to be bothered because I'm waiting for my turn to go up on that stage and have the president hand me my Master of Engineering degree. My last hurdle before real life begins, I already have a job lined up. I can't wait to get my degree and high tail it to California, where Hewlett-Packard is waiting to make my dreams come true.

And here it is! They call my name (with all three parts). I climb the steps and cross the stage. He hands me the padded certificate holder with the goods inside. I pause for the pic, smile graduationally, and in a flash I'm walking off the stage and returning to my seat.

As I settle back onto my preassigned griddle, I take a moment to reflect upon all the time and effort it took to get here. All the tradeoffs, the finagling, the sacrifices made. It's all led up to this. My one-year master's degree is finally in hand (two months before my 22nd birthday). I did it! I open the beautiful binder and feast my eyes upon... a letter from the financial aid office. WTF!

I don't know if I should raise my hand or just yell out. I'm dazed and confused, reeling from the sudden onset of graduatus interruptus. I'm not sure what to do. Perhaps I should read the letter.

Upon a quick perusal, the letter seems to indicate that my degree is being withheld until I make good on my outstanding student loans. I can understand the University wanting me to make good on my obligations before freeing the degree, nothing unreasonable there. The only problem I have with this whole situation is that I *don't have a student loan*! I'm here on a scholarship. In fact, I've never even communicated with Office of Financial Aid. I can't imagine how this could have happened. I've had issues with Tulane bureaucracy before, but they've really outdone themselves today. This must be rectified immediately. My only question: Should I wait until the end of the graduation ceremony or go visit them right now?

I opt for waiting it out, but as soon as that recessional forms I bust out and head straight for the financial aid people. I hand them the letter and, since I'm already dressed for it, demand my degree. They check it out and agree it's a mistake. Of course, no one knows how it could have happened. They'll be happy to correct it, but I'm already "in the system" so it will take time to resolve. Then they will simply mail the degree to me, if that's OK?

But it's *not* OK. I'm scheduled to leave for California, and I need the degree to claim my fabulous job. Besides, they screwed this up once already, why would I have confidence they will unscrew it in time? This was supposed to be one of those perfect, clock-stopping moments in life that I crave. I spent four years making this souffle, and when the timer finally chimes "ready," I open the oven to find it's fallen. My long-awaited entrée to adulthood and real life is hanging in the balance. Outside I'm furious, but inside I'm afraid…

Aaaaaand as I return to the present, I'm happy to find I'm not upset about the Hawaii trip surprise like I was about the graduation fiasco. But a tinge of the old ire remains. That's the way it is with emotional triggers. They transport you through time and space to a far less comfortable place. Decades from now, as a therapist, I'll be well versed in triggers and navigating the murky waters of reactivity. But here at Atari, I do not yet have that skillset, I just have the baggage. After the grueling workout of the last two games, my emotional state is quite raw. I'm only beginning to understand how raw.

And speaking of baggage, the next stop on my victory tour is a long-delayed trip back east to visit the family. At the beginning of the game, I mentioned to Ray that a visit home to see my parents was long overdue. I was making plans to go when E.T. raised his head and pressed me into service. Ray was sympathetic to my plight and assured me they would "fix that" for me after this project, which certainly wasn't going to be too much longer… right?

Though his intentions were golden, I'm not sure this was a beautiful gift. A great Zen master once said, "When you believe you have achieved tranquility and inner peace, go spend some time with your family." Ooooohhhhmmmmm My God! Just when I'm most vulnerable to getting my buttons pushed, I'm traveling to the factory where they were installed.

But I'm going just the same. I guess that's why they call it family obligations.

Good to his word, shortly after the delivery I was offered a ride on "The Hawker". This is the true corporate vehicle, a Gulfstream private jet. This is the plane Skip had complained was unavailable for our previous jaunt to see Spielberg.

I will be sharing the ride with Manny Gerard, his wife and their college-age daughter. Manny is the number two man at Warner Communications. He is principally responsible for Warner acquiring Atari. He is a rising star in the corporation… and will remain so for several more months at least.

My fabulous-flight experience begins days before takeoff.

Someone calls me to ask what I'd prefer to eat on the plane, lobster or sushi? It's the nicest in-flight menu I've ever heard. I say, "How about both?" They say, "Fine." Yes! Flying just keeps getting better and better these days.

Upon returning to the private plane terminal in San Jose airport, I can see The Hawker waiting on the tarmac. As I enter the cabin, it is clear this plane is significantly more luxurious than the Learjet which started this project. Manny and family board shortly thereafter and off we take.

The seats are super comfortable, and the sushi is delicious. The flight attendant is so gracious and accommodating. If the goal of this sort of extravagance is to impress people, then it's totally working on me.

I never spent much time with Manny before, but on first impression he struck me as a pleasant down-to-earth guy. We chatted a bit here and there. His wife and daughter were polite, but not cordial. Perhaps they are the seasoned travelers, and I am the wide-eyed newbie. I enjoy people watching, and sharing a private flight is an intimate experience. As time goes on, what jumps out at me is how this is such an extraordinary experience for me, yet to them it all seems so passé. What's it like to be used to this kind of lifestyle? And why is this so in-my-face every time I board a private jet?

None of this makes the trip any less delightful for me. We land at LaGuardia airport in New York, say our goodbyes and I will never see any of them (or the corporate jet) again. As I prepare to exit the plane, the attendant is kind enough to offer me a doggie bag containing the lobster leftovers. I graciously accept. It makes the limo ride back home to New Jersey that much sweeter. This is also the last time I'll see a limo for quite a while. It's like being in a wonderful dream just before waking up. Everything is perfect with no end in sight, and moments later it all disappears. Gone but never forgotten.

After surviving the visit home, I make my way back to California where I plan the third and final leg of the victory tour, Hawaii!

Mike Moone's assistant tells me to just book the trip, take the vacation, then turn in an expense report afterward and they will approve it. I've always wanted to go to Hawaii. Going on someone else's dime makes it even better. Condon, my project oversight wizard, has conjured up an excellent hotel in Maui. My bags are packed and off I go, not fully realizing all the baggage I'm carrying.

Ah, Hawaii. I always imagined Hawaii to be this amazing tropical paradise. But when I finally get there, it turns out the whole place is an amazing tropical paradise! I'd never been to a hotel where the front desk is nestled in a lush garden. Even when I'm indoors, I'm outdoors. The warm breeze feels perfect. Summer in Hawaii is ridiculously hot, yet this is the coolest place I've ever been.

OK, it's time to relax. I hang out by the pool, on the beach, on the lanai (Hawaiian porch), and at each of the many bars around the hotel. I play golf on the resort courses, drive to incredible sites and landmarks around the island. I stroll along the streets of Lahaina, stopping occasionally for pu-pus.

I'm ordering room service when I feel like it and hitting the restaurants when I don't. I'm doing whatever I want, whenever I want, so I should be having a pretty good time. The only problem is, I can't stop thinking about what's going on at Atari. I even call Jerome to see if anything's up. Naturally, nothing is up. The only news is: Some guy who is supposed to be on vacation is calling in to work.

I can't seem to relax. I'm in the perfect situation for relaxation and I can't quite get there. I even hang out under the famous Banyan tree in downtown

Lahaina, waiting for a local entrepreneur to offer me some pakalolo (that's Hawaiian for marijuana). I procure some Maui Wowie and hightail it back to the hotel. After smoking some, I can see why people rave about it. This is sensational pot. Unfortunately, even after repeated applications I'm still not quite at ease. I'm booked for two weeks at the hotel. After six days I can't take it anymore. It's time to go back. I love Hawaii. I know I will return to this incredible place, hopefully many times. But right now, I cannot deal with feeling so anxious in the most easy-going place I've ever been.

I hate to judge, but when you can't relax on an all-expense paid trip to a luxury resort in Hawaii, something is definitely wrong, even if I'm not ready to face it yet. What will it take to get my attention? Who knows? Another triggering event might do the trick. Fortunately, I won't have long to wait for that experiment.

Back at the office, I can't help noticing I'm having trouble settling down to work. My thoughts drift back to Hawaii. Why did I come back so early? When I was there, I felt the need to be here. Now that I'm here, it seems like I should be there. Mostly what I'm doing now is looking for distractions. Is it lunch time yet?

A most welcome diversion is on hand today. An Atari Alum Lunch is called in Mountain View at Frankie, Johnny & Luigi Too. It's a great Italian restaurant that always puts too much mozzarella on their pizzas, exactly as it should be. An alum lunch is where current Atari game makers get together with former Atari game makers (which means Activision and Imagic game makers). We are all seated around a long rectangular table, chatting enjoyably when the topic of E.T. comes up.

Apparently, word has spread that I did the game very quickly. Some are impressed, others are skeptical. Despite the fact I have witnesses who attest to the truth of it, some people are unwilling to believe it happened even after it had indeed happened. One such person sits across from me. Glancing up from their menu, they ask-tell me, "You borrowed code from other games, didn't you?"

"Nope. I wrote over 6K of new code. The rest was graphics."

"You didn't write all that in 5 weeks. No one writes code that fast."

And suddenly I'm 8 years old…

Staring out the huge windows of his third-grade classroom, which is how he spent eighty-five percent of his class time, young Howard could see it was a beautiful spring day. Precisely the kind of day that maximizes his frustration with this statutory incarceration adults call education.

After a long New Jersey winter, the biting cold was finally leaving the air. The walk to school was changing, it was no longer the arctic expedition he despised. It was becoming a jungle safari, involving treacherous river crossings that would place his schoolbooks and homework in great jeopardy. His daily excursions would be much more enjoyable from now on.

Having finished the math lesson for the day (the only part of school Howard enjoyed) it was time for penmanship exercises. The death knell. He glared at his pencil, and the pencil seemed to glare back. Howard hated pencils, but only students who demonstrate good penmanship are allowed to use pens. No matter how hard he tried, the script letters that Howard drew looked only vaguely like the perfectly formed characters adorning the "Good Lettering Strip." That strip spent the entire year sitting ominously above the blackboard like a vulture, an imposing monument to anal-retentives (a term he did not yet know, but thoroughly understood).

Howard knew all the letters and could use them to write and communicate. But the letters he made sure didn't look like the ones on the good lettering strip. In fact, nothing he ever wrote or drew looked at all like what he saw in his mind. Howard had always written this off to those magical "can't be explained" areas of life, much like when you order food in a restaurant and then it magically appears on your table. Consequently, the only thing the good lettering strip was useful for was gauging Howard's vertical leap. A purpose in which his teacher failed to see value. The strip also served as a constant reminder that he was hopelessly relegated to pencils until fourth grade. Smudge, smudge, smudge.

The teacher, Miss Reynold, the awe-inspiring object of Howard's first awakenings to sexual fantasy, centered herself in front of the class. Standing right under the perfectly written 'Mm' she said, "O.K. class, now we will do our penmanship exercises. Take out your pencils, or pens for those of you who have shown excellent penmanship," Howard groaned, "and William will pass out sheets of paper to everyone. Today we are going to write a poem about spring. When you get your sheet of paper, write your name in the upper right corner. Then under that write my name, Miss Reynold, using proper capitalization. Then write your poem. When you are done, bring it up to me and I will check it for you."

Howard really enjoyed writing poems because he knew lots of words, even if he couldn't spell them, and he enjoyed matching them up in rhymes. He wrote his poem, a cute little ditty about bells and birds, and he took it to the teacher for her inspection. He felt it was a good poem. Howard was frequently proud of his work.

As Miss Reynold read the poem to herself, Howard eyed her over and took himself to the limits of a third-grade sexual imagination. He felt a familiar, if unidentifiable, tingle.

Having finished the poem, Miss Reynold looked up at Howard. He was waiting to hear a typical "That's very nice Howard, but you are still crossing the case lines and not rounding out your 'o's and 'a's correctly." Instead he heard something that took away his tingle entirely.

"Where did you get this?" Miss Reynold demanded, looking him squarely in the eye. There was a quality of challenge in her voice that went way beyond questions of good penmanship.

"What?" replied Howard sheepishly.

"You didn't write this, where did you get this from?" she said very sternly, going into teach-you-a-lesson mode.

"I just wrote it in my seat. Right there." Pointing at his desk, he was completely bewildered. She asked him to write a poem and he did. He couldn't understand why authorship had become an issue, and even if it was, who cared? This was supposed to be penmanship.

"You didn't write this. What magazine did you get this from?"

"No magazine, I just wrote a poem." murmured Howard, who was scared and visibly shaken. He never liked being accused of things, particularly when he hadn't done them. And especially not by his favorite fantasy figure.

"Go back to your seat, we will talk about this later." Miss Reynold dismissed Howard by snatching the paper from the hands of the next child, slapping it down on her enormous desk and focusing on it intently. Howard could feel her switch from intense to ignore. He enjoyed some kinds of attention and was frightened by others. But no matter what kind of attention it was, Howard was acutely aware of it. And for right now, he was aware that there was no more attention left for him to enjoy here.

Howard walked back to his desk feeling an imaginary tail between his legs. He couldn't understand what had happened. He sat in what was suddenly a very lonely classroom. He reviewed the poem, the troublemaker. Fifteen minutes ago, Howard was a happy, albeit bored, young boy. Now this poem had turned him into a sad and defeated mess, robbed of his rhyming joy and his fantasies.

After some productive sulking, Howard thought about what had happened. The teacher said he hadn't written the poem, but he knew he had. She hadn't seen it anywhere else, or she wouldn't have asked him where he got it. So why didn't she think it was his? Howard thought on for a while and finally it dawned on him.

The poem was good. Very good. TOO good to have been written by him. Then he paused, it occurred to Howard that Miss Reynold might think he was too stupid to write a good poem. But she had always treated him as bright, with a teacher's concern. So, if she thought he was bright, and still thought the poem was beyond his ability, it had to be a really good poem. He felt reassured in his conclusion. In fact, Howard thought, it's sad she underestimates him that way. For the first time in a while, a smile began to grow on his face.

Miss Reynold would occasionally glance over at him while checking the other students' poems. Howard could hear her thinking, "What shall I do about this lying little plagiarist?" Each time she looked, her scowl got a little deeper because each time she looked, Howard's smile was a little broader.

OK, I'm back now. Sorry to have disappeared for a bit there, but that's what happens when we're emotionally triggered. In this case, I decided to switch to 3rd person for some healthy distance. That's the thing about traumatic memories, they're traumatic. I don't care to fully reexperience the entire ordeal. Nonetheless, this doubting lunchmate's comment sent me on a trip I didn't want to take.

When people doubt my authorship, their skepticism feels demeaning to me. It's as if they're saying I'm not good enough to have done what I did. This saddens me a little and pisses me off a lot.

It's possible I bring it on myself at times. I like to undersell things because I love the thrill of overdelivering. Unfortunately, this can invite others to underestimate me.

Maya Angelou said: "When someone shows you who they are, believe them." What I learned from these episodes is: When I show you who I am and you can't see it, that's more about you than me. Still, I get triggered, nonetheless. I'll have to work on that.

Sometimes I meet people who are excellent at setting me off. It's very difficult to be myself around those people. There are lots of extreme personalities around Atari, so naturally they evoke some rather extreme reactions.

Atari is a brilliant proving ground for a therapist-to-be.

But that -to-be is not to be for decades yet. Meanwhile, it's getting harder and harder to overlook the fact that little comments and moments are sending me reeling, even more so than usual. Is it time to seriously explore the question: What's going on with me?

Frayed so.

VOYAGE OF RECOVERY

Do you know what it's like to want something so much that it's more important than taking care of yourself? No matter what it costs to continue, it's still a happy trade-off. That's what making video games at Atari is like for me. There is literally nothing I'd rather do and there is no limit to which I will not push myself to achieve it. It's an addiction. And, as is frequently the case with addictions, there comes a time when I hit my credit limit and payment is due.

I couldn't let go in Hawaii. Not even for one week. I'm so addicted to Atari that I needed a break from my break.

Obviously, I'm suffering from a fractured relaxation bone. Though, to be fair, that bone was already calcium deficient. Relaxation is a skill I never really mastered, or even bachelored. I spend my time focused on what needs to happen next, and I'm usually busy nudging *everything* I can in that direction. Easing off and letting things flow is not part of this lifestyle. What if they don't go the right way? What if I could have made the difference and didn't? Responsibility weighs heavily at the intersection of need and obsession.

But the broken bone is merely a symptom. Bones break under stress. What is my stressor? What sinister force could turn my reward into a punishment? Future therapist Howard believes I'm worried it might all disappear. This whole beautiful life, my wonderful gravy train might dry up and fall off the tracks. I have no idea how prescient this will turn out to be.

I'm also skillfully overlooking the fact that my "beautiful life" has me so tweaked out I'm unable to enjoy paradise. But hey, this period of my life is not about being reasonable. Don't forget, when you expect things to make sense, you're losing touch with Atari.

Regardless of my willingness to accept it, I am burnt out beyond recognition. If there's a Phoenix waiting to rise from these ashes, it's still sitting in an airport in Arizona, unticketed.

It does seem a period of recovery is necessary. How will I recover? How long will it take? They say the first step is admitting there's a problem. So, however long it may take, it's a cinch that clock hasn't started running yet.

I'm already feeling pressure to do something special. I don't want to break my streak of successes. I just finished a 5-week development. How do I follow that?

It's not like I have no ideas. I designed a Yars' Revenge sequel I believe will play very nicely. I was going to start it just before they told me to go interview with Spielberg for Raiders of the Lost Ark. Then I was going to start it when Ray Kassar called with another priority project. Now I am completely free to do whatever I want, so it's time for Yars the 2nd, right? No problem.

Except this next game needs to be amazing, and my design uses the paddle controller which has limited distribution. I'm worried this might limit sales. Although there's a funny inside joke about paddle controllers lacking potential, I'm not laughing. I fear my best current idea isn't good enough. But if not that, what?

I experiment with a variety of play concepts. I put up lots of screens, but nothing measures up. I'm supposed to be able to make magic, yet it's painfully clear I've got nothing right now. It's hard to unleash your creativity when you have to make a hit.

Each day I come in to work and head directly to the game room for my morning triathlon, one of the few rituals I'm maintaining. My morning triathlon consists of breaking 100,000 points on Defender, Robotron and Millipede. This takes anywhere from fifteen minutes to two hours, depending on the day. The triathlon is my daily workout routine, but I'm not feeling the usual rush. These are brilliant games. But when I play them now, there's an audio loop running in my head: "This good or better. My next game has to be this good or better." Then I go sit at my desk and try to come up with something fresh, dynamic and unique. I'm finding it hard to sit still for too long though.

Lately I'm better at talking myself into difficult situations than I am at designing myself out of them. Devoid of my typical enthusiasm, I'm just marking time, hoping against hope that something will pop.

As September comes to a close, I find myself slogging into the fall of

1982. After a brilliant ride, my outlook is bleak. I'm feeling thunderous echoes of two years ago, the autumn of my discontent.

It is a dark time.

HERE COMES THE SUN

Through October and well into November my burnout gradually eases, giving way to an aimless malaise.

Fortunately, people are willing to grant me some space during this time. It's too bad I can't schedule my recovery like we did the E.T. game, but recovery takes as long as it takes. Some things do, regardless of desired timing.

[NOTE to the Schedule-Conscious Reader: This book is taking a lot longer to write than I wanted, or predicted, or promised. Longer than my revised projections, and longer than my subsequently updated projections. But I refuse to make the same mistake on this book that I did on the video game which serves as its inspiration. I choose to prioritize product quality ahead of deadlines. As the reader, you are the ultimate arbiter of quality. I hope this choice serves you well.]

As the days grow shorter I am getting steadily better, finally returning to my normally atypical self. And just in time for...

December of 1982 is an amazing month, with some wonderful high notes, a harbinger of doom and some life lessons I will carry forever. It's practically a whole book in itself. There's even a bridge in it, as in: I'm going to "a bridge" it since I don't want to write another book when I haven't finished this one yet.

You know the picture of me holding the British Airways ticket in my hand? Jerome snaps this picture on one of the first days of December. This ticket takes me to London, where I'm reunited with my grandboss George and a very cool guy from international sales, Steve Race. I'm here to attend the British premiere of the E.T. movie and do some TV interviews to promote

DEC '82 -
LONDON PREMIER of ET

the video game. Now *that's* what I call a business trip. I must say, every time I fly on behalf of Atari it's a wonderfully worthwhile experience.

At the movie premiere, our Atari delegation pulls up to the red carpet and we stroll through the waiting gauntlet of paparazzi. With shutters clicking and flashes flashing, we are serenaded with chants of "Who's that?" and "Are they anybody?" I smile and wave and keep walking.

Inside, an usher escorts us to primo seats near the center of the theatre. After a while, things settle down and it looks like the movie is about to start. Then a swell of music rises, and so does everyone in the theatre. I follow suit, then follow the eyeline of everyone around me. Entering through the main lobby doors are Prince Charles, Lady Di and Steven Spielberg. They are escorted to our general vicinity and eventually settle in their seats, three rows directly behind us. This is exciting! British royalty will watch me watch E.T.

London is great. George and I have a great time at Windsor castle. Steve takes me to Harrods, which is most enjoyable. Then my TV interviews get cancelled. So, after a few jet-lagged but lovely days in the UK I head back to California.

Upon my return to the office, I find the latest Billboard Magazine Top-selling Video Games chart posted on the bulletin board. Both E.T. and Raiders of the Lost Ark are on it. I now have two games in the top 10. Welcome home!

Days later, fourth quarter royalty checks come out and mine is bigger than the secret E.T. bonus I received four months earlier. Hello!

It was a tough lead up to it, but December of 1982 is going fabulously for me. I have two best sellers, I'm getting big checks and attending movie premieres overseas. No one is using the phrase "worst game of all time" yet. In fact, I'm getting nothing but positive feedback for completing the

E.T. game. Sure, I have some anxiety about what I'll do next, but I know it's going to be a video game for Atari. There's nothing else I'd rather do. Life is beautiful.

Look how things have changed. A mere 24 months earlier I was sitting in a performance evaluation at Hewlett-Packard, deeply depressed. I was thinking: What the hell is happening here? What am I going to do about it? I didn't have any answers. I've had many upsets in my life but only a few depressions. I don't get depressed over adversity, bad luck or bad situations. I get depressed when I don't like where I am and can't see a path out. I've changed goals many times in my life, but I nearly always have one. When I haven't the slightest idea what to do next, that's when the darkness falls. Two years ago I was caught in the grips of my first real depression. Look at me now!

This is proof positive: No matter what's going on, no matter how bleak things may seem, I'm never more than two years away from a dramatically better life. This doesn't mean things *must* change, of course. Stagnation is a game anyone can play. But it is incredible how much things *can* change in just 24 months! This lesson will get me through many difficult times in my life, and I will see it come true in a huge way at least once more. Interestingly, that will be shortly after my third major depression, many years from now. Bob knows, and soon you will too.

I also learned to recognize and honor my need to create, to produce tangible results. There may be dry spells here and there, but I will always have a next-creative-endeavor in my plans.

Another valuable lesson I received is to honor the artist in me. I grew up in a family that respects art, but not artists. Being an artist is a crazy thing to do. I was raised to believe a person must have a trade. In other words, be a professional. I work hard to validate my technical side, but Atari helps validate my artistic side. This will continue to make a huge difference in my life.

My choice to pursue technology opened lots of opportunities for me, but Atari is opening me up to myself. It's the ultimate educational opportunity and I hope I'm paying attention.

On the whole, December of 1982 has been very, very good to me.

But it's not all roses. There is one other development of note this month, one that foretells of ominous clouds on the horizon and some freezing rain already falling. Apparently, while I was enjoying my royal screening, Warner Communications put out a press release announcing poor results for this financial quarter. The reason for the bad news is a precipitous drop in Atari's video game sales. In fact, the results are so bad that Warner's stock price tumbles dramatically. This could make for an unhappy New Year for Warner execs and Atari officers. Fortunately, I'm not holding much Warner stock. It's still a jolly good time to be Howard.

[NOTE to the Reader Wondering About the Writer Writing: Since there's still some space left on this page, I thought I'd mention that I'm having some drinks while writing this. I've always heard the tradition of writers who drink as they write. They hang out in famous bars (whose fame derives mainly from the writers who drank there). They down glass after glass and produce the world's most venerated literature. It occurs to me that, since I'm getting kind of buzzed, I may have some real potential as a writer. This is spirited logic.]

CHAPTER 15
PATH CONNECTED

Looking out over the vastness of the desert, the phrase that comes to mind is: The Sands of Time.

The desert is amazing in that something so huge is made up of such small parts. It's a collection of tiny individual grains which combine to make an enormous whole. Sand is fascinating to me, it's a solid that flows like a liquid. Ironically, the way to stop sand from flowing is by adding liquid. I enjoy thinking of beaches as little deserts that hang out with large bodies of water. But I digress…

Let's get back to the sands of time. One place where flowing sand becomes significant is an hourglass. The hourglass is nature's version of a traffic jam. I turn it over and every grain in the top is trying to commute to the bottom. After a while they all get through. You turn it over again for the evening commute and the same bunch of sand moves back to the other end. However, if I trace the path of each grain of sand, I see no two journeys are ever exactly the same. They go in a different order and hang out with different neighbors on each trip. Some grains are together frequently and some never meet. And then there are ones that connect once and don't reconnect until many trips later. People are like grains of sand in an hourglass. We are each on our journey and we're surrounded by countless others on theirs.

I have always thought of video games as a broadcast medium. There are aspects of making a video game that go way beyond the basic work experience. I'm talking about the possibilities, the unanticipated increments that accrue from bringing entertainment to people you don't even know. Through my games I've touched millions of lives. It's done at a distance, but done nonetheless. Seeing all the people here in Alamogordo, I realize that every one of them has been drawn here by a seed I planted over 30 years ago. The thing is, when I plant a seed remotely in someone's life, I never know if their path will cross my own and what fruit it may bear.

Today, hundreds of paths are converging in Alamogordo. This isn't a garbage dump, it's a crossroads in time and space.

So many hugs and handshakes. Some are first time meetings, some are reunions, but everyone here has one thing in common: They feel a connection with the E.T. video game, and by extension, me.

They are all here, proudly displaying their E.T. regalia. Some purchased, some homemade, but every piece is a precious talisman brought to the altar for sanctification (the ritual for which appears to be getting my autograph on it, whatever it may be). Over time, I've learned events like this demand preparation. That's why I bring a cache of indelible markers, in high-contrast colors.

MIKE MIKA & HSW.

I have never met most of today's attendees. There are a few happy exceptions however, and one remarkable exception to the exception.

Zak Penn and the film crew are here. They are all recent additions to my life, and welcome ones. As a video producer myself, I always enjoy watching a production crew in action, and these people are good! It's a pleasure to watch their smooth execution and high level of professionalism. And it's especially nice to see them in a space larger than my dink-ominium.

Mike Mika is here, too. Mike is an important part of the video game world and our paths have crossed before, starting many years (and several Yars) ago. Mike and I first talked in the mid-1990's when he was developing a Game Boy version of Yars' Revenge. He is a committed and talented game maker. We became friends and our paths have run parallel ever since, occasionally crossing. I'm enjoying spending time with Mike again, and today he introduces me to a good friend of his, Ernie Cline.

You can tell Ernie is here by the ANORAK license plates on the DeLorean parked at the dig site. The gull wing doors went up once the storm subsided, revealing a life-size E.T. sitting in the passenger seat. The doors will go back down when the wind returns to prevent the DeLorean from becoming a sandbox with wheels.

Ernie wrote one of my favorite novels: Ready Player One, but that's not why he's here. He came because Ernie loves classic games and he cannot resist this kind of opportunity. Ernie is, and I say this with reverence, a nerd's nerd. That's not how he puts it. Ernie describes himself as an enthusiast and a gentleman adventurer. The thing I like about him is he lives up to these titles Ernest-ly.

Upon meeting Ernie, he presents me with a real Indiana Jones-style side bag containing an autographed copy of Ready Player One, and a full-sized bullwhip. This is handy since my old Atari whip is long past its last crack, though I wonder if (and how) he knew.

ERNIE CLINE, ZAK PENN & HSW.

I am touched by his thoughtful generosity, and awestruck at meeting an international bestselling author. Meanwhile, he is telling me how much he enjoys and admires my games. This is awesome. The longest day of my life is sporting some truly amazing moments.

I'm hanging out with a bunch of talented and accomplished people, all because I made some games over three decades ago which touched their lives. A delightful dividend from my karmic lottery ticket.

This is the magic of Atari: I get the opportunity to connect with millions of people. As they continue their journeys and I continue mine, no one knows where or when these paths may cross. And it all springs from doing something I dearly love: Making a video game.

This is what's going on above the surface. But this is an Atari story, which means the likelihood of unseen undercurrents flowing beneath the surface is quite high.

One thing I notice is Ernie and Zak tend to step away from the group in down moments. They huddle, speaking in muted hushed tones. I don't want to intrude but I wonder what's up.

They are obviously working on this movie together, but I will come to find out this isn't the subject of their discrete discussions. It turns out they are also collaborating on another project, a secret the world won't hear about for three more years. They are co-writing the screenplay for a Ready Player One movie. And who signed on to direct this film? None other than Steven Spielberg. When anything E.T. related is happening, the aura of Spielberg is never far behind.

Meanwhile, back on the surface, a commotion is going on and people are gathering around the back of one of the production trucks. Zak is standing on the raised lift, using it as a makeshift podium. He has an announcement to make. As the crowd settles, he holds up a small black piece of plastic and tells us that one of the fans has just found this. It's a joystick cover for an Atari VCS controller. It's the first nugget of this excavation. The crowd cheers and the anticipation builds. We are getting closer to paydirt.

The sands of time are flowing before our eyes. Sometimes gathering to hide things. Sometimes clearing to reveal them.

The dig continues, now with even greater enthusiasm. Moving tons of sand, chasing after sacred artifacts and antiquities. Delving ever deeper, trying to find the buried treasure which spawned so much lore over the years. Then it hits me: We may be looking for the E.T. game, but we are living Raiders of the Lost Ark.

THE FORGOTTEN GAME

I did four games at Atari. I'm known mainly for three games since my fourth, Saboteur, wouldn't be released for 20 more years (making it the *longest* game development in history). My first game, Yars' Revenge, is my glory. My third, E.T., is my infamy. My fourth, Saboteur, is my unfinished symphony. But my second game, Raiders of the Lost Ark is the forgotten game.

E.T. gets lots of attention, and Yars gets its fair share of press. Saboteur ultimately got excellent reviews, but Saboteur can't be the forgotten game because nobody ever heard of it in the first place.

Somehow, my Raiders of the Lost Ark game got lost in the shuffle. I've done over a hundred interviews about my games, but Raiders has only come up two or three times. In my trio of million sellers, Raiders is the overlooked middle child.

Check out this table:

GAME	DEV TIME	VOLUME OF ATTENTION	MY ...
E.T.*	**5 weeks**	!!!!!!!!!!!!!!!!!!!!!!!!!!!	3rd
Yars*	5 months	!!!!!!!!!!!!!!!!!!!	1st
Raiders*	10 months	!!!!!	2nd
Saboteur	14 months	!	4th

(* indicates over one million units sold)

It seems the more time I spend on a game, the less attention it receives, which is a rather odd feeling for me.

[NOTE to the Sales Number Nerd: Each of my first three games ended up selling well over a million units. Even after subtracting returns, E.T. still crossed 1.5 million in sales. Since Saboteur was never released back then, I can say honestly that every game I released for Atari was a million seller. I don't believe anyone else (who did more than one game) can make this claim.]

I've mentioned Raiders of the Lost Ark a few times so far, but even in this book it's been an afterthought. Art imitates life. Let's talk about my second game…

In the trajectory of my Atari career, Raiders was an abrupt turn. Whereas Yars was a game for the eyes and ears, Raiders was a game for the mind. Instead of a sensory overload, I wanted Raiders to feel like the biggest game any VCS player could imagine. It was my most ambitious undertaking (except for the others).

Raiders started amid the turmoil around releasing Yars, continued through one bonus plan introduction, an attempt to leave and go to another company, a second bonus plan introduction, and ultimately ended right at the beginning of the ET project. No wonder I walked around with a bullwhip.

As a developer, Raiders of the Lost Ark is interesting because it's a switch from action to adventure. In my opinion, there's a huge difference between making action games and adventure games; The maker of an action game can have the same play experience as any other player. With adventure games, this is impossible because adventure games are all about secrets and puzzles. If I already know the secrets, how can I assess the level of challenge? I can't. My intellectual virginity is lost, and I can never go back.

Raiders revealed my preference for designing action games. I like immediate feedback. I can feel an action game firsthand and tune it appropriately. Adventure game design frustrates me because I can't get that player experience. I must either tune it blind or watch others play which requires a steady supply of fresh subjects. What a pain.

And speaking of pain, I felt tremendous pressure to succeed on Raiders of the Lost Ark. This was for two reasons, both self-imposed. First, Raiders was going to be Atari's (and the world's) second adventure-style video game. The first one, Adventure by Warren Robinett, is a genre-defining masterwork. Warren blew players' minds, creating a killer app for video games. I needed to make a significant contribution to the genre or be dismissed as a cheap knock-off. In other words, if I'm going to stand on his shoulders, the view better be spectacular.

The second source of pressure stems from the success of Yars' Revenge. Doing Yars was extremely validating and gratifying. I felt good about it, but it doesn't take long for insecurity to start seeping back in. I'd ask myself: Am I a one-hit wonder? Is that all I've got? Was I just lucky? These questions haunted me throughout this development. I didn't become truly confident as a game designer and programmer until I finished Raiders. And just as I did, a stubby legged alien came knocking on my door.

Then there's the pain my quirky nature causes others. Believe it or not, I have a somewhat contrarian nature. It's not about arguing, it's about avoiding unpleasant surprises. As a problem solver at heart, I am frequently looking for the trap, the loophole and the exception. I'm also seeking the elegant or alternative solution. The issue is I frequently find them.

Consequently, I tend to approach things in a different way. I rarely fit in because I'm rarely doing things the way other people expect them done. My goal is not to do things the way I'm "supposed to," I just want to do them well. Full disclosure: I really enjoy violating expectations in a productive way. Tell me there's a standard way to do something, and I will try to show you a non-standard improvement. This makes me an innovator. In fact, given two equally productive approaches, I will choose the path less traveled. Flexible results-oriented people find me entertaining and useful, but I tend to irritate the crap out of more rigid types.

Innovators like to bust myths and they frequently have trouble following rules (especially rules that seem unduly restrictive or unnecessary). At

Atari, I was truly an innovator, and there was one rule I violated for sure…

Nolan's Law, handed down by Nolan Bushnell himself, was the fundamental video game design principle at Atari. It states:

> "The best games are **easy to learn** and **difficult to master**. They should reward the first quarter and the hundredth."

This is a great rule, for coin-ops. I don't think it applies to home games. I prefer "difficult to learn, difficult to master." Here's why:

If I can ask more of a player initially, I can make a deeper game that provides longer playability and greater satisfaction. I made some games which required the player to read the manual, or as we say in tech, RTFM. Coin-op players might walk away after two or three quarters, but home gamers pay a full bag of quarters right up front. I can ask more of the player because they're already motivated to get their money's worth.

Raiders busted other conventions as well. For instance, I used two controllers for one player. Using 2 joysticks in Raiders was an odd thing to do, but it created the possibility of complex inventory control and added a new dimension of game play. I intentionally did it in a way that would be problematic at first, thus signaling the need to check the manual at the outset. This was a hassle for some people, but many worked through it to enjoy something fresh and new. Was that a good choice? I don't know. It did sell over a million, but it didn't sell three or four million. I often wish life had do-overs.

I also opted for a graphical score. I liked this idea since you can see how you are progressing in the game. But it really lacked score clarity, and the player can never actually reach the ark. This is a decision I regret, and it's the first thing I'd change on Raiders of the Lost Ark. I would still use the graphic display, but I would also show the numerical score and the potential score. To all the players who found this frustrating in the game, let me just say this: Sorry, my bad.

One other thing worth noting about Raiders is how it engenders one of my personal life philosophies. As I've said, video games frequently exemplify the beliefs and values of their designers, and my second game is no exception. In Raiders of the Lost Ark, there are places where you seem to get trapped or stuck. The truth is: Players are *never* stuck in this game. One of my basic tenets in life: When it feels like life has me cornered, there is always a way out. I may not see it, but I fundamentally believe it is there just the same. This also shapes my therapeutic approach. I don't tell people what to do, I help open their eyes to fresh opportunities and possibilities. This enables them to make better choices and create positive change. When I feel stuck, I'm usually just unaware of alternatives. That's why it's so important to stretch myself when exploring the world around me. Test the boundaries to discover which are walls and which are doorways. In Raiders of the Lost Ark, as in life, you are never truly stuck.

Another "one other thing" worth mentioning is that Raiders was the game for which I was best outfitted. Walking around with my whip and hat, cracking the whip in the hallways, made working on this game more fun for me.

I did fight with marketing toward the end of the game. They wanted to put a lot of the secrets in the manual so people would be able to play the game. I didn't want to do that since adventure games need to present a real challenge to players and handing out the secrets seemed like the opposite of that. It's important to remember this was all pre-internet. There were no cheat guides or video run-throughs to teach the average player how to advance. In the end, marketing won out and I turned over all the secrets. I want to be crystal clear about this: Marketing was absolutely right! I thank you merry marketeers for protecting me from myself in this case. Even in the toughest of relationships, listening can still help things work out just right.

Raiders was ten months of intense development and developments. There were distractions and conflicts and long hours and hard choices all along the way. There are certainly some oddities in Raiders of the Lost Ark, but there is nothing fundamentally wrong with the game. I would not venture to say the same thing about E.T.

There is definitely something wrong with *that* game…

WHAT'S WRONG WITH THE E.T. GAME?

Overall, I believe the E.T. video game is a solid playable game done in record time. That said, I do not think the game is flawless by any means. It has plenty of real problems, chief among them being the fact that many people don't like the game. I never argue with player opinions. If someone doesn't like a game, that's the truth and I don't try to convince them otherwise. If you've played the game, you likely have your own ideas about it.

There are some typical complaints about the game, mostly around falling into pits, getting out of pits and figuring out what to do in the game aside from falling into and getting out of the pits. But these only scratch the surface. There are far deeper issues arising from my five-week foray. If you'd like to criticize my effort more completely and accurately, please allow me to arm you with my top answers to the question: "What's wrong with the E.T. video game?"

In my mind, the single biggest problem is committing the fundamental sin of video games: Thou shalt not disorient the player. Please do not confuse this with frustration. Disorientation and frustration are very different emotions. Frustration is knowing what I'm trying to do but feeling unable to do it. Disorientation is feeling completely lost as to where I am or what's happening. Frustration is seeing the cookie but I can't reach it. Disorientation is stepping into the kitchen for a cookie and finding I'm in the garage.

Frustration is essential to video games because frustration is what makes goal achievement feel rewarding. A game without any frustration is not a game, it's a chore. Taking out the garbage is not a game. Sneaking the garbage out past the hungry gargoyle and through the secret teleporter, only releasing it in the correct sector of the sacred dump of the ages… that's a game. Obstacles and opponents are the kinds of things which provide frustration (and ultimately satisfaction) in a game. But a player should never be disoriented. In a good game, the player is frequently failing but always understands why they are failing so they can practice, improve and ultimately succeed. E.T. has too many moments when the player is simply disoriented and that is a major flaw in the game. The truth is: Some elements which I hoped would constitute "game play" challenges really don't land that way for most players. And to those boys, girls, women and men, I say: Sorry about that.

Also, like Raiders of the Lost Ark, E.T. violates the "easy to learn" clause of Nolan's Law. I did that on Raiders for the sake of game depth. On E.T. it was a matter of expediency. It's quicker to explain things in the manual than to develop a non-trivial self-explanatory game. Honing and simplifying are expensive processes. It's a sentiment as old as revision itself. As Mark Twain put it, "I apologize for such a long letter - I didn't have time to write a short one."

But these are conceptual transgressions. Circumstances played a key role in compromising quality as well...

The most succinct yet comprehensive explanation of this is: E.T. was released at First Playable.

Remember "First Playable"? After the design is adopted, the next major milestone in a video game development is the point when all basic game elements are present and all rules of play are implemented. No bells or whistles, just the bare essentials. It's the first time the game can be experienced largely as the design intended. This is known as First Playable. It occurs (hopefully) somewhere around 30-50% into the planned schedule.

First Playable is really the starting point for a game because it's the first time you get to feel the game, to see what works and what doesn't. Then you begin the most important part of the development, the Tuning Phase. Ideally, the majority of development time is spent adjusting and improving the game. With E.T., it was clear there would not be time to tune or refine the game. Simply making it to First Playable in 5 weeks would be a considerable achievement.

Another problem was the lack of Rumination Time. On creative projects it's very important to have the opportunity to make it, get sick of it, step away from it, clear it from your mind and then come back to it with fresh eyes. I never had a chance to take any real break from E.T. This put inordinate pressure on the original design to be perfect. Which leads to the next problem with the game…

E.T. achieved virtually 100% of its original design concept. Ordinarily, this does not sound like a problem.

"Hey there, how'd that project go?"

"We achieved 100% of our original concept."

"Oooh, sorry to hear that."

This conversation sounds ridiculous, but the fact is the final versions of most successful products are quite different from their original designs. Why? Because they get *better*. For a variety of good reasons, original designs are merely launching points, not true destinations. As I work my way from vision to reality, I stumble on fresh approaches and insights. New possibilities arise. Good developers shoot for a final delivery which is significantly better than the initial concept. All my other games benefited from this trajectory. E.T. did not. There simply wasn't time.

I think the best way to illustrate the value of tuning and rumination time is an actual demonstration of the creative process. And I can't think of anyone better to exemplify this than one of the great creative minds of the last millennium: René Descartes, the famous French philosopher and mathematician of the 1600's.

[NOTE to the Title Conscious Reader: Philosophers are people whose job consists of coming up with sparkling nuggets of insight and wisdom that convert nicely to marketing slogans. This is not to be confused with advertising executives. Ad execs frequently employ philosophers, but strictly on a work-for-hire basis. As for Mathematicians, please do not be fooled by titles. "Mathematician" is simply old-school slang for nerd.]

And now, here is the story of one of Descartes' more noted (and now storied) developments…

René Descartes made many contributions to humanity, perhaps his most famous is the phrase: Cogito Ergo Sum, which is Latin for "I think, therefore I am." (back in the 17th century, nothing counts unless it's in Latin). We're all familiar with this quote (at least I think we are, therefore we must be). But precious few know the story of how René arrived at this gem. It wasn't some random flash of insight, it evolved over several rounds of an ongoing creative process.

Descartes set out upon a quest for pithy yet profound philosophical insights, as philosophers are wont to do. While having a snack one day, an interesting thought occurred to René: Dooreetoe Ergo Numyo, which is Latin for "I eat junk food, therefore I am delicious." This never really caught on, because even early adopters in the 1600s were not quite ready for the concept of junk food. He decided to step back and ruminate for a while.

After a raucous Halloween party, René was inspired. Scrapping the junk food concept, he decided to go in a more esoteric direction with: Incognito Ergo Summa, which is Latin for "I disguise, therefore I am fabulous." René felt the seed of something special in this idea, but he wasn't quite sure how to germinate it. He decided to hire a market research consultant to check it out. The consumer testing showed significant resistance to phrases which begin and end with vowels. The unfavorables were overwhelming, with the Extremely Opposed and Somewhat Opposed combining for over 73% of the sample. Respondents also felt (by a 2-1 margin) there were too many N's.

René was crestfallen, but in time he started thinking about the results. He thought and he thought, until finally he started to think about the fact that he was thinking. Over and over he'd ask himself, "Am I thinking?" And each time the reply came back "I am!" René lost sight of his product and was obsessing over his process. He couldn't let it go. When his incessant perseverating became too much, he decided to pay a visit to his friend and fellow creator, Auguste Rodin, who also lived in France, just a couple of centuries over.

René spent many hours sharing his perturbations about thinking and thought. This inspired Rodin to create his famous sculpture, The Scream. Apparently, Descartes exceeded Rodin's tolerance as well as his own.

Upon returning home in time, René went straight to his lab and began tuning his "Incognito Ergo Summa" project. Recalling the market research, he disemvoweled the head and tail. Next, he removed the N's. Finally, he dumped the second 'M' at the end which now seemed superfluous. Et voilà! Cogito Ergo Sum was born. Descartes' contribution launched modern rationalism and validated consumer testing methodology for centuries to come. But wait, there's more...

Being a person who despises waste, René wondered what he might do with the discarded letters. He threw away the vowels, of course, because in France vowels are everywhere. But what to do with the rest? He took the leftover consonants and used them to create a tiny confection which he dubbed "M&Ns". Sadly, this never got off the ground because he couldn't write the recipe in Latin. He was forced to settle solely for the philosophical contribution. C'est la vie!

This is possibly (though admittedly implausibly) a true story. Nevertheless, the fact remains: Time for tuning and rumination are critical to the success of a creative product. On E.T., that time simply wasn't available. Of course, in development, everything is a tradeoff. In this case, the lack of critical development time was traded for an abundance of critical feedback.

Another issue with E.T. was timing... historic timing. I did E.T. before there was an internet. This means there was no "drop", there was no instantaneous customer feedback, and once feedback was received, there was no chance to improve and update the product. What goes out initially is all there will ever be. This increases the pressure to get everything right the first time. Think about this: With the internet, I could have started the game on July 27th, worked until October 31st (tripling the schedule) and still released the game on November 1st!

Then there was my desire/need to innovate. I didn't want to just finish a game in record time, I wanted to make a contribution as well. That's not unique to E.T., I want to do that on all my games.

E.T. introduced the first 3D world (players maneuver around a cube), also E.T. features the concept of location sensitive powers. What you can do depends upon where you are standing. It's also an entirely non-violent video game, unusual for the time. I'm not sure if these things helped or hindered the product, since these choices were typically about making it faster to code. But despite a ridiculously short schedule, it wasn't enough to finish the game, I still needed to break new ground. It never occurred to me that 30 years later I might be watching this game break *old* ground... in Alamogordo, New Mexico.

In the beginning, I was a tad miffed that Spielberg wanted me to do a Pac-Man knock off. But I must admit, looking back over the whole thing, he just might have been on to something.

HOW WOULD I FIX E.T.?

Interviewers who are familiar with the game often ask me what I would change about the E.T. game. Believe me, I've thought about this over time. There are so many ways this can go. I think the best way to view it is by answering another question: How much extra time do I have to make the changes?

Let's say I had one extra day to work on the game. Given what I know now, what would I do differently?

With one day, I would make a few tweaks to gain absolution from committing the sin of disorienting the player. Here are the changes I could make in one day to fix this:

The player would only fall into a pit by placing most of E.T. over the center of a pit. It would require a much more affirmative action than just touching any part of the player to any part of the pit. This would reduce the "unhappy surprise" aspect of falling in.

Next, I would make it impossible for a player exiting a pit to fall back in. Upon return they would be placed next to the pit with motion disabled until the joystick is released. Now they can only go back into the pit intentionally.

The last thing I would do is change the arrow-powerups so when you jump to another screen you always land in a safe place. The same would go for walking onto another screen. I would make sure the player can never accidentally fall into a pit.

I believe I could make these changes in one day, given the time and desire. I had hoped the challenging navigation would enhance the gameplay. What I hadn't counted on was players' expectation of reality. They didn't see abstract graphics interacting with one another, they saw characters walking around and expected them to behave realistically. I shouldn't fall into a pit because my head touches it, only my feet should make me fall in. I'd never encountered this expectation in a game before. I attribute this to the high-quality Jerome brought to the graphics in the game.

If I had an extra week, I'd have done the above and also added three extra features. First, I'd add a visual transition for falling into and coming out of the pits. This would make it less jarring and should keep the player oriented to the game situation.

I would also add a radar screen so the player could see if any humans were approaching on adjacent screens.

But the coolest thing I might have done is to address my Nolan's Law violation. I would add verbal instructions, so the player knew at each point what to do next. Either "Build Phone," "Call Home," or "Meet Ship." It could have been the first home game with on-screen directions. That would've been innovative, helpful and doable too.

And if I had an extra month or two? Well… then I would redesign the game, shifting the goals, mechanics and layout significantly.

But honestly, by the time E.T. was done, so was I. There might have been time to make a couple of tweaks and change a few things, but by then I was so full of E.T., I couldn't take another byte. I didn't want to hear or do anything more with it. Besides, if I'd have gone and made some fixes the game might have risen to the level of "not bad," thus obscuring its place in history. Think of all the pop culture and media content that would have been lost forever.

Worse yet, we wouldn't be here right now, and I'd hate to have missed this opportunity to chat with you so enjoyably.

CHAPTER 16
FIDDLING WHILE ROMS BURN

1983 is an interesting year for Atari. It starts in recovery, progresses to intensive care, then ultimately winds up on life support. Of course, that doesn't mean we have to deal with it.

Thanks to an unhealthy dose of denial, we in engineering are feeling invincible while the ground crumbles beneath us. We are insulated both inside and out, blissfully ensconced in our bubble.

BMOBS, though powerful, cannot stabilize our foundation as we conspicuously ignore the foreshocks of the coming reality quake.

WE DON'T BLAME YOU

One thing we didn't really get a lot of in engineering was feedback. We got a sense of how sales were going by our royalty checks, but very little direct feedback. We received schedule requests and license notifications, and that was about it.

This makes it particularly strange when one of the marketeers comes up to me in February of 1983 and says, "You know Howard, we don't blame you. You really came through for us."

I smile, thank him for the kind words and assure him it is my pleasure to be of service. As he walks away, I am thinking: What the hell is he talking about?

This happens another time or two with people from marketing or management. I'm always OK with not being blamed for something, but it would be nice to know what I didn't do. But I never ask. I guess it's a why-spoil-the-moment sort of thing.

Remember, it's been six months since I finished E.T. and virtually all the feedback so far has been positive. That little alien is finally out of my head.

Now I'm totally focused on creating my next game. Still, it's an odd thing to hear... repeatedly.

I LOVE IT WHEN A GAME COMES TOGETHER

What's my next game? I was finally coming up with an answer.

It isn't like nothing is happening. I'm developing my next hit. I blow off my Yars' Revenge sequel because it requires the paddle controller. That could limit sales, something my BMOBS inflated ego will not tolerate. I need to follow three hits with a fourth and I'm playing the odds... sort of. The thing is, I have a design I believe in for the paddle controller, I do not yet have an alternative I like.

E.T. was the shortest development ever done on the VCS. My next game, Saboteur, will be the longest. Measured from start to release, E.T. was 36 days and Saboteur will be more than 20 *years*. I believe that's another record. This game won't be released until 2004, two decades after I finish working on it.

Saboteur is my return to straight action gaming, my preferred genre both as player and creator. After months of recovery from burnout and a variety of failed gameplay experiments, I was finally zeroing in on something I believed would be worth doing. A multi-screen twitch-action sensory extravaganza with never-before-seen game mechanics. It took a while to cobble it together, and then a while more for Atari to figure out what to do with it.

GETTING INTO CHARACTER FOR THE A-TEAM CONVERSION OF SABOTEUR.

Why? Because during the development, Atari negotiates a license for the A-Team television show. They tell me to switch this game from the original Saboteur to the A-Team license. This means Jerome has to create a whole new set of graphics and I have to rework several of the scenarios to make sense in terms of the license context. Doing E.T. taught me that a game is not just a game. The game's theme will greatly influence players' expectations of what reality ought to look like in this low-resolution imaginary world.

Ordinarily this is fine, I'm good with the A-Team, I even got a Mr. T headcover. But one day marketing decided it wasn't that great a license and called it off. Then the game started flip-flopping between Saboteur and A-Team. It got to the point where we maintained two versions just in case. It went back and forth a few times, and when it looked like it was time to make a choice and finalize one version for release, our high-resolution real world exploded and the game was shelved.

Like Raiders of the Lost Ark, this game endured a series of tumultuous environment changes. During this development there were two building moves, two CEO changes, many internal Atari quakes and one major earthquake (magnitude 6.2). But alas, one thing it will not do is be released while I'm still here. Twenty years later, the Saboteur version will be resurrected, finalized, packaged and ultimately released (along with Yars' Revenge) on the Atari Flashback system.

Saboteur is my unfinished symphony. You can tell because it's the only one of my games with no signatures. It does feature some Yars however, which had graduated from hidden Easter egg status to explicit game element.

To paraphrase the A-Team's leader, John "Hannibal" Smith: I love it when a release comes together.

NO 2ND ACT

There were two kinds of people at Atari, those who hoped the VCS would last forever and those who wanted to see gaming take the next step forward with a hot new console. And then there was the third kind of two kinds of people, those who lost bladder control over the idea of cutting our own throat by releasing a new machine before they sucked every last drop out of the VCS money-river.

Everyone knew there would have to be a next generation, but the real trick to releasing a new console is timing. Ideally, just as your sales are falling off the old console, the new one steps in and the baton is passed. The problem is: That's never happened before, so we have no idea where the sweet spot is. It's the curse of the first product lifecycle. The decision is delayed as long as possible because the way management sees it, introducing a new product is undercutting ourselves. Let's just keep cashing in. Unfortunately, this mentality leaves them totally unprepared to deal with what happens at the end of a first successful product life cycle.

Ultimately it is decided we need to get something out into the market, and that's when they discover the huge lag time from drawing board to production with a new console. Uh oh.

In true E.T. style, Atari snapped into reaction and rushed to market with too little too late. The Atari 5200 made it out just in time for the 1982 Christmas market. Advertised as the latest in video game technology, it was really the existing home computer, repackaged with a new controller that had its own problems. Almost everyone was disappointed. Engineering wanted better tech. Marketing wanted more game titles for release. Consumers were upset to learn about a new technical term: Backwards Compatibility. No one can play their existing VCS games on the new 5200 console. UH OH!

Engineering is split also. There is an underlying tension between programmers on the 5200 vs the VCS. For the near term, 5200 royalty potential is far less. Developers pay a price for working on the better hardware. Sure, that will change over time. In a couple of years, the 5200 sales will dwarf the VCS, everyone knows that. After all, people aren't going to stop playing games. Right?

When you work in the video game industry there are always new toys and systems to explore. The real challenge lies in knowing when you're the one getting played!

HUBRIS AND HEMLOCK

Atari was not a nurturing environment, but it did cultivate a number of things. 1983 saw a bumper crop of attitudes, egos and entitlement. Not for everyone by any means, but some individuals really distinguished themselves. When dramatic success comes breezing in through the window, complacency and entitlement gather like dust. Little by little, the layers accumulate until they cover everything, dulling the colors and distorting the original beauty. It manifests in different ways for different people. There was no greater demonstration than one afternoon in the parking lot behind the building.

This is the same parking lot where our remote-control race cars crashed and smashed into each other in a violent ritual, purging our pent-up fears and frustrations. It was our high-tech demolition derby, and it served us well. But today our frustration has a different voice, and that voice is standing next to me, making a joint's ruby tip glow brightly.

Several of us are clustered together, billowing smoke, but the voice stands out from the others, still tanned from an all-expense paid tropical vacation in a premiere resort (it's not me)(this time). It is also my understanding they recently received a favorable financial concession from the company in an effort to assuage them. They finish off an impressive toke and courteously pass the joint along. Then, in that funny voice of someone who must share a thought immediately but doesn't want to exhale, they say, "You know, I don't have to take this anymore. They can't keep treating us this way." And I remember thinking: OMG, people are losing their minds!

It is amazing how easy it is to adapt up to better circumstances, and then to believe they must get better still. To do quality creative work takes confidence and ego. It's a necessary but not sufficient condition. In other words, I can have plenty of confidence and ego and still produce garbage, but without them I won't get anywhere.

Hubris was Atari's greatest resource, and ultimately its poison. Too often, bold innovative spirit devolved into brash unmitigated gall.

In fairness, the voice was frequently dissatisfied. When I first got to Atari, I couldn't believe how amazingly free and open the environment was, yet there were people (like Alligator-clip Jim) walking around saying, "This place used to be so great. I can't believe what's happening." I eventually got to the point of saying it too. The difference being, of course, that it was true when I said it.

THE STORM IS BREWING

Atari achieved a significant milestone, becoming the fastest growing company in American business history. When it comes to climbing the ladder of success, there's a classic Hollywood quote:

> "Be nice to people on your way up because you'll meet them again on your way down."
> WILSON MIZNER, PLAYWRIGHT & ENTREPRENEUR

This, however, was not Atari's credo. The Atari way was more like: When you're leading the race to the top of the heap, dump your garbage on the other climbers. After all, these are the spoils of victory and that's just how power players play.

This philosophy, applied consistently over time, helped Atari establish another significant milestone. They became the fastest falling company in American business history.

Easy come, easier go. Have you ever noticed how escalators tend to run in pairs? For every escalator going up there is usually another next to it going down. Sometimes it felt like we were riding both simultaneously, which is an odd feeling indeed. It's OK at first, but it gets increasingly uncomfortable as my feet move farther apart.

While things were still ramping up for me and E.T., there were little gremlins at play in the industry, driving the down-scalator. As late '82 rolled around, things started to appear less rosy in the executive suites. We started to see signals that the executives were reading some rather toxic tea leaves.

Of course, even if the signs are clear, it doesn't mean you have to read them. It's like Ayn Rand said: "You can ignore reality, but you cannot ignore the consequences of ignoring reality."

We were like people partying in Pompei, marveling at the smoke coming out of the mountain. See how the gods favor us with beautiful sunsets these days, they must be happy. Party on! There were no geologists to tell these partying Pompeiians otherwise.

CHAPTER 17
WHO'LL STOP THE RAIN

The desert is lovely at the moment, nothing but big sky and an endless, clear horizon. We're just passing time now, waiting to find out if the big yellow monsters will dredge up anything worthwhile.

Looking west toward White Sands National Park, it's all good… until I look a little closer. Through the hazy heat-distorted air rising from the desert floor, what I thought was a clear horizon is in reality a big white wall, and it's heading this way. The white sand of White Sands is fueling the next environmental onslaught. Everything is fine right now, but it's only a matter of time…

I'm surrounded by all kinds of people here in the desert, from heavy machine operators to light snack vendors, from local civic leaders to travelling classic gamers. But the most surprising to me are the anthropologists. I don't think I've seen an anthropologist since Margaret Mead back in college. Once I think about it though, I realize they are essential to this effort.

Anthropologists are the detectives of lost and ancient cultures. We need them on this mission to uncover remains of the lost culture of Atari, and perhaps discover reasons for its demise. When it comes to corporate culture, Atari is a rich vein indeed.

Culture is what made Atari… and ultimately what destroyed it, taking the entire industry down in the process.

I arrived at Atari amidst a massive cultural migration from Nolan Bushnell to Ray Kassar. It was an interesting and ominous time.

Like the approaching wall of white sand, the great video game crash is coming. Two things make it inevitable: The presence of culture and the absence of knowledge.

A CULTURED APPROACH

Once Warner takes over, they install Ray Kassar as CEO. Slowly but surely, things begin to change. There is nothing wrong with change. Improvements are changes. But this particular transition is a shift from the Nolan regime (which favors people who make things) to the Ray regime (which favors people who make money from people who make things). This is an epic cultural shift.

In some ways the entire story is told on their first encounter. When Ray Kassar arrives for his first day clad in executive standard attire (cloth and cologne), he is introduced to Nolan Bushnell sporting Atari chic (t-shirt and jeans). In fact, Ray said in a 2011 interview that the t-shirt Nolan was wearing featured a vulgar declaration of Nolan's desire for physical intimacy. Nolan denies this unequivocally, and I have never heard Ray's assertion corroborated by any Atari veteran. Gauging by my experience of Nolan, I believe he would aim higher. Even if a loving nature was the sentiment, his t-shirt would couch it in a far more articulate framing.

This transition is a tale of two cultures that couldn't be more different. It's a transition from production-aware management to production-unaware management.

Nolan culture was Creativity culture. It was about rock stars and making better games. It was an engaged culture, connected to the source of the product.

Ray culture was Meet-the-Schedule culture. It was about pre-sold licenses and getting things out on time. It was an aloof culture, disconnected from the source of the product.

Nolan nurtured programmers. He was a product-first kind of guy, and that's how he managed. This inspired creative people. Programmers loved Nolan and would follow him anywhere (and did so for decades in companies long after Atari). This is in sharp contrast to Ray Kassar's seeming indifference to programmers, and unwillingness to recognize their value. I believe it's more accurate to say he was unaware of their value, but to many it felt more malicious. Like the time he told David Crane & Crew they are nothing but a bunch of towel designers. This attitude inspired enmity amongst those he needed most to augment and maintain his success.

Atari needs innovators, but innovators come with factory installed boundary issues. Boundary issues are good for thinking outside the box, but bad for keeping your arms and legs inside the vehicle. Nolan had a knack for creating free space around the box to accommodate innovators. Ray and company are trying to put innovators back *in* the box. This trick never works.

Programmers can be managed but take care not to tweak them. Most programmers have a penchant for two things: Resentment & Revenge. Managers who inspire these skillsets rarely last.

Ultimately, Ray's attitude led to the exodus of key programmers which led to the formation of a competitor which led to the discovery that Atari had neglected to legally lock the VCS which led to the revelation of immense video game profits which led to a frenzy of frothing financiers which led to a glut of new company formations which led to a glut of crappy games saturating the market which led to the collapse of the industry which led to this run on sentence. Perhaps a little more time and attention to the programmers initially might have paid off better in the medium to long term.

There are always signposts along the transitional highway, the trick is to recognize them. And as we all know, these signs are far easier to read in retrospect.

One of the more profound markers of cultural transition (and decay) is how brainstorming is increasingly replaced by blamestorming. The grand offsite brainstormings disappear, though massive salesforce extravaganzas remain. We spend more time explaining why things aren't done and less time figuring out what to do next.

Another tradition that withers and dies is the Friday afternoon party. These parties weren't only about drinking and debauchery, they were the primary mode of interdepartmental cross-pollination. Lots of productive ideas came from casual chats between colleagues whose schedules wouldn't overlap otherwise. These parties were a staple of the Nolan regime. As they disappear, the interdepartmental contact goes with them. This is a big loss for everybody.

Gone are the days when engineers have a showdown on the coin-op game to settle who gets to make the home version. That assignment method reflects a culture focused on doing the best game.

Another place the transition is painfully evident is release criteria. When is a game done? It used to be the game had to be approved by the maker/players, now the schedule seems to dominate. If the game has enough pre-sold market potential and quality assurance says it's sufficiently bug free, out it goes. That's a significant change.

I'm often asked: When did things start to unravel? It was when the goal shifted from making good games to meeting schedules. Or if not unravelling, it is at least snagged and frayed.

Another factor contributing greatly to Atari's demise was the transition from Meritocracy to Egotocracy.

Every company, whether it employs thousands or just one, must accomplish three basic functions. "Engineering" must create & maintain the product, "sales & marketing" must advertise, sell and distribute the product, and "management" must keep the company productive and working effectively.

In reality, major successes occur when all three are done well, because each relies on the others. Of course, reality is not the same as truth. We all share one reality, but everyone's got their own truth.

In a meritocracy, we share a common truth: Let's make our best contribution, upholding our commitment to the others.

In an egotocracy however, engineering, marketing and management each have their own truth:

Engineering truth is: Without us no one goes anywhere because we make the product. If we make a better product, you'll have better sales so leave us alone and let us give you the great product everyone is waiting for.

Marketing truth is: We can sell anything. It doesn't matter what it is. The only thing we can't sell is nothing, so stop playing with it and give it to us. We will sell the hell out of it and everybody wins.

Management truth is: You know, these engineering and marketing people are all insane. They're so full of themselves, none of them realize how important we are. Without us, nothing would happen. We just need to make sure their attitudes don't end up costing us money, status, productivity and ultimately profits.

Each believes they are the key to success, as long as the others don't blow it. This is BMOBS at its finest. As a therapist, I see this line of thinking in relationships as well. It's amazing how people who need to work together can spend so much time blaming others. Blaming in a partnership is like saying: "Your side of the boat is sinking."

When people think a core department in their company is unnecessary, it's a sure sign of corporate pathology. A well-run company considers everyone's input and maintains a staff worth listening to.

Management in Ray's Atari was surprisingly removed from our product. You don't expect a CEO to be a full-fledged developer, but to make good decisions you would expect some awareness of what's involved in creating the product and what makes it good.

Consider this: At a time when games take six months for design and implementation, the biggest movie license deal in history was given 36 *hours* for design and less than five *weeks* for implementation. They took more time negotiating the rights than they allowed for making the game. And they didn't consult the people who must deliver the game until after the deal was signed. This should give you some insight into how things go in Ray's Atari.

I'm guilty as well. When the licensing addicts finally went shopping for prescriptions, I was the only doctor who said: "Absolutely I can." Another BMOBS triumph.

The E.T. game did not cause the video game crash. It is, however, symptomatic of the thinking that caused the crash. I see it as the crowning achievement of this mindset.

Atari's cultural transition produced a dramatic shift both in corporate worldview and corporate results. Under Nolan it wasn't about making sense, it was about making fun and it ended up making a lot of money. Under Ray it still wasn't about making sense, it was about making money and it did at first, but it ended up making a disaster.

Nolan was trying to *learn* how to run this new kind of business. Ray was hired to *know* how to run the business, whether he did or not. I believe he (and his Warner overseers) believed he did.

When we switch from trying-to-learn to assuming-we-know, we pay a toll. This is the cost of BMOBS. It pulls us away from productive curiosity and pushes us into ill-informed judgment.

Cultural shift was one of two major tributaries to the great video game crash. Their confluence formed a mighty river which flooded the valley and devastated the community.

NOBODY KNOWS ANYTHING

William Goldman was a twice-Oscar'd screenwriter and a perceptive member of the Hollywood movie making community. He wrote several all-time classic films, but his most famous line wasn't from a movie, it was about the movie industry. It's William Goldman's fundamental rule of Hollywood: Nobody Knows Anything.

Anyone with an intimate partner understands this principle. When you're guessing what others will love, hits are hard to predict, and misses can be *very* expensive.

I believe Goldman's rule also applies to the dawning of video games as well. In fact, Nobody Knows Anything is the other major tributary to the great video game crash.

You may recall many pages ago when I mentioned the simplest explanation for the crash: It was the first product life cycle.

Let's unpack that now.

Atari was the first major player in the shiny new field of techno-tainment. It was the first time a video game system left the atmosphere. The most important thing to remember about unexplored territory is: It's unexplored. Bold adventurers may return with incredible treasures, but only if they return.

As with any new endeavor, there are many questions waiting to be answered. Chief among them: What questions should I be asking about this new endeavor? I do not yet know what I don't know. That's the challenge of new ventures, it's a high stakes race to see if I can learn the essentials before my ignorance kills me.

Here's one question: What business are we in? Atari masquerades as a technology firm, but it's product is clearly entertainment. That's Warner's specialty, so it's reasonable to assume they understand the business. But it wasn't that simple.

Atari was a hybrid. A new kind of company creating the new medium of interactive entertainment. The VCS was the first child to truly thrive in this new world. What do we feed it? When do we change it? How stinky are the diapers? Nervous parents look for answers, but it's the first major product life cycle in a new industry, which means nobody knows anything.

OK, we know one thing: There will be problems. In fact, there will be three kinds of problems.

There are old problems like corporate structure, product manufacture and distribution. Marketing must convince consumers a luxury item is essential. There will be politics, power struggles and competing agendas. There is nothing new about BMOBS either.

There are also new problems, like managing a whole new world. Got new approaches? Then there's the rise of the Intellectual Blue Collar. Don't worry, I'll explain that shortly. (Whoa! Easy there, Bob.)

Then there's the apex issue, this is State-of-the-Art. When it's broke, nobody knows how to fix it. When it works, nobody's sure what to do with it. A solution in search of a problem.

Nobody knows how the first product life cycle is supposed to go. My mother knows how it's going to go, though: "You make all your mistakes on the first one."

Warner buys the company and they start to wear Nolan Bushnell down. When it's time to install their own management they pick Ray Kassar, a person who doesn't seem to understand technology or entertainment. He does, however, understand corporate power games.

Ray will have his glory, then everything will crash and burn. Would Nolan have created the same situation? Probably not.

But Nolan sold the company rather than attempt to drive it to new heights. He sold it for less than $30 million just a few years before it would reach a billion. Why? Because he couldn't take Atari public at the time and didn't have the money to launch his revolutionary brainchild. Such is the quandary of the innovator. The only way to ensure a healthy birth was to bring in help. Enter Warner Communications Inc. Of course, this invites the too-many-cooks conundrum.

But the birth was successful and off it took. It was a beautiful baby, but it carried the one congenital defect which, more than anything else, opened the door for the great video game crash of the early '80s: They forgot to lock the console.

Recall the analogy to music on vinyl records. There's the media (the record) and then there's the player (turntable). Manufacturers make players, producers create media. Turntables don't care who made a record. If it fits on the turntable and the groove holds the needle, it plays. Initially, manufacturers did both since no one else knew how to make a record. Eventually, people figured out how to make records on their own. Soon lots of producers were happily making records and selling them. Manufacturers didn't like this, it cut into their profits from music sales. However, since turntables cannot choose which records they will or won't play, the music industry grew to accommodate both manufacturers and producers.

The first home video game systems went the same way. There's the media (cartridge) and the player (console). The big difference is: A game console *can* decide which cartridges it will or won't play *if* you set it up to do so. That's called locking the console. Every modern video game console does this, which allows manufacturers (Sony, Microsoft, Nintendo) to control (and get paid for) *every* game released on their system. By the way, "modern console" means after the VCS. If Atari had "locked" the VCS, the flush-clogging glut of crap could not have happened. Alas, they didn't.

Back then it may not have seemed necessary for two reasons. First, video game tech is way more complex than record player tech. Second, there weren't many systems out there so profit potential was quite limited… until the VCS took off!

Even then, it would take someone with an ex-employee's knowledge to understand how to approach it. And financiers would have to learn of the huge profit potential. Ex-employees might know that too.

But where is anyone going to find tech savvy ex-Atarians? Ray will take care of that soon enough. And when other competitors/producers start making games for the VCS, Ray answers by beating the hell out of them with lawyers. When the smoke clears, it turns out any bozo who can make a game cartridge is free to sell it on the unlocked VCS. And every bozo does. In a world without standards, average game quality tanks abysmally.

Atari was all about innovation. Our technical innovations grabbed people by their imaginations and paid them off in reality. We benefitted greatly from those. Unfortunately, our legal innovations benefited the competition and killed us. This wasn't stupidity, it's just that nobody knew. That's the cost of the first product life cycle.

Atari made plenty of mistakes, but locking the console would absolve many sins. It would not, however, cover this next one…

There are two kinds of people who drive companies: There are Innovators who create explosions, resulting in successful new products or craters. Then there are Maintainers who create stability, resulting in smooth and productive functioning.

There is no better or worse between innovators and maintainers. Both bring value and both make mistakes, but their skillsets and blind spots fall in different areas. Either can be effective *as long as* they are in phase with the company.

Unfortunately for Atari, at a time when innovation was needed to extend the life of the VCS and create a next generation console, Warner switched from an innovator to a maintainer. Who knew?

This is all very conceptual. Let's look at how it played out…

When Warner takes over, they go looking for a new leader to take this simmering sauce to a boil. Enter Ray Kassar.

Ray comes from a big multinational textile firm with a traditional corporate management profile. Warner likes that because big corporate execs tend to want other big corporate execs to run companies for them. That makes sense. Unfortunately, it turns out the textile management approach doesn't fit the fledgling techno-tainment industry. It's a nonsensical act, but it's done from a sensible perspective. Classical management making a classic mistake.

They install Ray Kassar as CEO, and soon thereafter things really take off. Tens of millions become hundreds of millions. Now everyone is a genius. Ray is a genius for creating this obscene cash bounty. Warner execs are geniuses for buying Atari and putting Ray in charge. Those are the circumstances, but what really happened? There are two basic ways to look at this:

One is: Ray is a brilliant manager who saw a nicely working plane sitting in the shop. Using some spare parts he found lying around, he upgraded the plane to a rocket and launched it.

The other is: Nolan built an incredibly powerful rocket. He set it on the launching pad, got it all warmed up and ready to go… and then he sold it to Warner. They changed management and Ray Kassar came in and watched the rocket take off.

Which is it? Who cares? We appear to be making good decisions and we're doing really well. Obviously, we rock! It's like they say: Success has a thousand parents, but failure is an orphan.

But it makes a big difference which one is true. Think about this:

Techno-tainment is the wild frontier of business. Warner Executives try to apply conventional reasoning to a very unconventional situation. They do what has worked for them in the past because people go with their strengths, especially when they get paid a lot of money to get results. That makes sense.

But you know what happens when you expect things to make sense around Atari…

The crash is definitely coming. So, what are they missing?

Atari is not a classical business, but Ray is a classical manager.

The org chart is the classical manager's model of the company, it informs their view on employee value. Since the industrial revolution, assembly line workers have occupied the org chart basement.

Modern software companies bend this model, and creative technical environments turn it on its head. They also introduce a new kind of org chart member: The Intellectual Blue-Collar worker. They sit at the bottom of the org chart, but they have substantial control over the product. They are often smarter or more insightful than their org chart superiors, which can create ego conflicts and power struggles with insecure managers.

Nolan Bushnell understood this, which made him a good person to drive this new kind of organization. Then he sold it to Warner, who didn't understand. They hired Ray.

Ray saw programmers as the bottom of the org chart and dealt with them that way… at first. Ray was confident in his abilities but, like most of us, he was unaware of his blind spots. He didn't know what he didn't know. For instance, he didn't know that he didn't know programmers are well versed in resentment and revenge. Here's how you know he didn't know:

Look at the towel designer incident. When David Crane & Crew (soon to be known as Activision) started bubbling up, Atari made a business decision. They thought it unwise to take a handful of above-average programmers who were making around 30K and raise them to 60K or 70K and throw them some bonuses. Know this: An average game programmer can generate millions in profits once per year.

This is obviously a ridiculous financial decision, but it wasn't about economics. It was about ego-nomics. This is where hubris raises its ugly head in the Atari story. Here we have a classical manager facing a mutinous situation. You can't have line workers telling you how to run the business and extorting money from you, just replace them and go on. There was no concept of game makers as craftspeople or artists, they were simply tools of production and replaceable as such.

Ray didn't understand what he had, and because of that he drew a hard line. When the golden geese came asking for more feed, Ray said, "No. Get out of here! Who do you think you are?"

He could just as easily have said, "You know what, we will work out an increase in compensation that'll make you happy. Now go make some more games." But maybe he couldn't have said that.

Maybe this man in this moment found it impossible to say that. He couldn't see that spending a little money on programmers now could save loads of money in competition later. He didn't know, and in this case his ignorance was extraordinarily expensive.

The formation of Activision is the intellectual blue-collar version of a strike. Only instead of a traditional strike where people don't go to work, they strike out on their own and start another business. Launching a new software venture is easier than starting a traditional manufacturing effort. This fact leaves poorly run tech companies far more vulnerable to competition from former employees.

Initially, it looked like Ray knew how to blow the doors to success right off their hinges. But when you open the floodgates you have to remember to protect the city. Ray lacked the toolset for that.

He let the cat get out of the bag and competitors popped up like weeds (some actually doing good product). Ray's solution was to sue them or buy them. He tried desperately to fix the problems (or at least relieve some of the pressure), but it was too little too late. Cracks kept appearing and precious profit was leaking. It was one of the great ironies of Atari that Ray Kassar became the person charged with plugging holes in the dike.

The storm is coming, and those who see it are all asking the same question: Who'll stop the rain?

SYMPATHY FOR THE DEVIL

I'm fortunate to tell this story as a therapist because it helps me find empathy for all the players, including Ray. If I told it merely as an ex-Atari employee or an engineer, it wouldn't be the whole picture.

Curiosity and Judgment are two things that come up a lot in therapy. Managing them well enables me to interpret my world accurately.

Case in point: How do you think I feel about Ray Kassar?

So far, I've reported my interactions with Ray, and I've shared my interpretations of some of his deeds, but I don't believe I've discussed my feelings about him.

By now, you might think I don't like him very much. This is judgment. It's based on something (the story so far), but it's only an assumption. On the other hand, you could ask me how I feel about Ray Kassar. This is curiosity. Asking questions invites better information. Of course, if you ask me right now, I probably won't answer unless you happen to be here with me. Given this is unlikely, let's pretend for the moment that you are so I can.

I always liked Ray Kassar, at least I enjoyed the Ray I knew. This is neither defending nor condoning Ray. I can like someone and still disapprove of or criticize their actions.

I want to highlight a few things about Ray and his circumstances...

This is not the story of how Ray was an idiot. Ray was not an idiot. He was a seasoned business executive who was unfamiliar with the product/industry he was hired to manage. When you don't understand a situation, it's easy to do the wrong thing for the right reasons. And it's always hard to understand never-before-seen scenarios.

Nolan was better equipped to manage this kind of endeavor, but he didn't fully grasp the nuances of console development either. That's part of the price you pay for being first. Nolan totally whiffed on the legal side of things, but no one will realize this until later, long after he sells Atari. Ray's feedback will be more immediate.

Ray has no experience managing entertainment or hi-tech and they put him in charge of a breakthrough entertainment technology company in a brand-new industry. In short, Ray is the wrong guy in the right place. Nonetheless, his early results are wildly successful!

Everything is coming up roses. Warner execs are laying huge bonuses on

Ray. He takes the bonuses and buys into the idea he's an amazing manager, doing great things for Atari. Classic BMOBS.

What would you do if you accidentally launched a rocket resulting in people heaping praise and rewards on you? Turn it down? Admit it wasn't really your doing? It takes a lot of character to turn down a million-dollar bonus. More likely you do what Ray did.

But when things started to break at Atari, it was painfully clear Ray did not know how to fix them.

Nolan knew how to treat creatives in a way that motivated teamwork and productivity. Ray did it in a way that bred resentment. This ended up costing a ton of money and some key defections that snowballed into colossal problems. But I don't think it was about money for Ray, it felt more like a need for people to respect his structure and know their place. Though he came by it honestly, Ray was in the wrong industry for that mindset. His tragic misread of the situation led to a series of unfortunate decisions and ultimately his downfall.

Ray Kassar was no angel, and he definitely made his own bed. Still, I have empathy for him. He was esteemed as the genius who made Atari fly incredibly high, only to discover he had no answers when the sun started melting his wings.

THE DECLINE AND FALL OF ATARI

In early 1982, Atari felt like a solid foundation that would last forever. Cracks started showing over the summer of '82 and by late autumn there were gaping holes. During '83 it became crystal clear that no one knew how to fix them. The foundation crumbled, shattering the dream. Things fell apart so quickly, it was astounding.

Cultural shift, Nobody Knows Anything and BMOBS joined forces to send the most successful company in American history (and it's industry) into an epic death spiral. But that's very theoretical. Here's my insider view of how it actually played out…

One significant indicator of the leaking foundation is a stock sale on the part of Ray Kassar and another senior Atari exec, who reportedly dump a non-trivial amount of Warner stock (Warner being the parent company of Atari). People notice this sale. Not people at Atari so much, but rather people at the Securities and Exchange Commission. They notice it because of the timing. The stock is sold *very* shortly before Warner announces its financial results, and the numbers are bad. Very bad!

The implications of a situation like this are many, here's the main one: It raises many pairs of eyebrows on the Warner board. Not because of the stock sale (that's just profiteering, which is pretty much their whole raison d'etre). What elevates eyebrows is the delivery of poor financial results, the only real corporate faux pas. This precipitates a brief inquisition into the matter. After all, Ray is their guy, a seasoned exec from a traditional corporate giant. They put him in place and tremendous success followed. He'll know how to fix things, right?

Well, after discussion and reflection it appears not. Apparently, the Warner board members are not familiar with the definition of State-of-the-Art. They are, however, intimately familiar with the meaning of a falling stock price. This leads to Ray's dismissal in the spring of '83. A strong move and a harbinger of… what?

Things start falling apart. Atari's early release of the bad numbers in late '82 (undercutting Imagic's IPO) really changes the mood of the industry. And what about the mood in Atari engineering? Is it raining on our fabulous parade? Not a bit of it. OK, let's just say I can see the clouds are starting to form but I'm still focusing on the silver linings. Never underestimate the power of denial.

The Warner board, however, can't afford denial. They need to act. How will they fix this situation? In early summer of '83 they decide to bring in Jim Morgan, yet another seasoned exec from a traditional corporate giant.

Somehow, the people at Warner just can't seem to grasp the idea that classical managers from huge monolithic companies may not be well "suited" to run a dynamic technology company in the fledgling medium of interactive television.

Perhaps they can only see as far as themselves. They are big corporate execs who have done very well, so maybe this is the only kind of person they can trust to take over. I'm just speculating of course. It would take someone with a background in business, technology and psychology to figure this out.

In his favor, Jim Morgan does have a new approach to the problem. True to business chic of the early '80s, he brings the corporate-raider mentality: If a company isn't making money, it's because we're paying too much in salaries. After looking things over for a couple of months, he starts reducing the corporate payroll. By early 1984, the total employment of Atari has shrunk from well over 10,000 down to 2,000 people. His tenure is short-lived though, he disappears when his new approach fails to stem the tide of red ink.

After going oh-for-two with imported "experts", Warner is finally waking up to the fact that big execs from huge companies are ill-equipped to save this hi-tech hot potato called Atari.

What now? Atari, the flagship of Warner's success, is hemorrhaging cash and dragging the entire empire down with them. Their two best shots at saving things through management machinations have done little more than prove out State-of-the-Art: It was broken, and no one knew how to fix it. Well, there's an old business saying I just made up: "If you can't save 'em, sell 'em!" In desperation, Warner sets out in search of a buyer.

And they find Jack Tramiel. Jack is not your typical corporate executive. He is an interesting guy with a remarkable background. After surviving Auschwitz, Jack left his native Poland and came to America. He opened a small business selling typewriters and parlayed it into Commodore Computers, a billion-dollar company. Then, in the tradition of many successful entrepreneurs, the company he launched and nurtured from little more than a dream had grown to the point of no longer holding his interest. Squabbles and disagreements have reduced it from an occupation to a mere job. It's time to find a new challenge, so Jack resigns from Commodore. Now he's contemplating his next corporate move, and Atari looks ripe for the plucking.

Warner is overjoyed to find Jack. Atari is principally composed of three divisions: Home Games, Home Computer and Arcade Games. Arcade games is the only one still making money consistently. The other two are ruining Warner's entire financial picture, which is no small accomplishment. Warner off-loads the Home Games and Home Computer divisions to the Tramiels for little more than a stock position in Jack's new company, just to get them off the books.

It's important to note that Jack Tramiel isn't corporate chic, he is hardnosed and all about business. He embodies another tradition of successful entrepreneurs, he knows how things need to be and he is going to make damn sure they go that way. The first thing he does is have a round of interviews with the current developers. He makes a couple of things crystal clear to me: What the new development environment will look like, and what my home life *should* look like. It is an interesting interview to say the least.

Sitting in the familiar office with a brand-new face behind the desk, Jack asks me a variety of questions. None of them has to do with technology since that isn't Jack's thing, but it isn't all about business either. He asks me if I'm married. I tell him I am, and that my wife works. He explains to me how I'm off base there. My wife should be at home, waiting with my slippers when I return from work. That's actually what he tells me. I'd heard about his background and I can appreciate where he's coming from. Different values but similar determination. Jack is old school about business, and even more hardcore about family.

In short order, the Tramiels take Atari from 2,000 people down to 200 people. I'm still here. I'm not nearly as enthusiastic about being here as I used to be, but I am still here.

I really get the feeling Jack and his three sons came to town on a mission, to create a product so good it would put Commodore out of business. I think the game they want to make is "Jack's Revenge!" It certainly isn't "game development", that's abundantly clear. As he lays out the plan, I can see there is no place for me.

I gave that interview some serious thought. I respect Jack immensely as an achiever, a pariah and a holocaust survivor. But I know I can't work for him, certainly not the way I'm used to working. Atari created a monster in me that all my future managers will have to deal with in one way or another.

I passed the point of "it isn't like it used to be" a while ago. That, in itself, isn't so bad. After all, Atari was so great it could fall quite a bit and still remain miles above the alternatives. What's hard for me is seeing hope devolve into floundering desperation. Watching brainstorms become blamestorms. Seeing my gravy train jumping the tracks and careening down the walls of the canyon. A majestic company is losing its character, and ultimately its mind. When the Tramiels appear, any lingering dream of returning to classic Atari culture is extinguished.

I can feel the deflation. Hollowed out shells of our former selves, we roam the halls aimlessly, waiting for the other slipper to drop. We know something dramatically different is coming but we don't quite know what. One thing is certain, this is the end of an era. Over the years, it was not unusual to occasionally think we might not have a job tomorrow, but it was a different thing entirely to think there may not be an Atari.

In due course, the word comes down and it's time to decide: Will we stay on in the new world or take a lay-off package? From the specifics of the deal, it's clear game development is not a serious consideration. Jack is obviously planning to make a home computer, since those are the only real positions available. I've done operating systems before and qualify to stay on. But I realize joining this new Atari is essentially returning to Hewlett-Packard, and I just can't do that. I opt for the lay-off package and head for the hills.

SAYONARATARI

Taking in the great expanses of New Mexico I must concede, the desert is a beautiful place. Of course, that beauty is based on a contingency. I can appreciate the beauty as long as I know I can turn around and walk into an air-conditioned enclosure with food and water waiting. Were that not the case, staring out over an endless desolate parched desert would inspire a different reaction entirely.

When I think of being stranded in a place devoid of everything I need to thrive, I'm reminded of one of the saddest moments of my life. The day I left Atari…

I pick up my black coffee mug with the gold Atari logo emblazoned on both sides and place it gently in the box. This cup will be gracious enough to hold my pens for decades to come. But it's early in our relationship so here I am, stuffing this iconic tchotchke in a cliché cardboard box. It looks comfortable, nestled securely in the center of my coiled Raiders of the Lost Ark bullwhip. My Romper Room diploma is there too, standing guard over its siblings in the roofless cardboard enclosure. This ever-vigilant diploma has witnessed it all.

Though freshly wiped, I see specks of white powder and fuzzy green bits trapped between the glass and the lip of the frame; Remnants of brainstorms gone by, dreams shared and plans made. So many intensely fine moments, how can they possibly fit in a box?

I scan the room one last time, thoroughly inventorying memories while checking for overlooked stragglers. How many millions of lives were touched by events hosted in this space? I thank my desk, pick up the box and head out the door. Dead man walking.

I can't hear my footsteps because that damn box will not shut up. It's piercing my denial and sounding the end of my all-too-brief stay in the era of wonder and true satisfaction.

There will be other jobs, other opportunities, even other game companies. But there will never be another Atari. The excesses of childhood don't always play out so well in adult-land. It's as true in business as it is in life.

Most people spend their time at work waiting to go on vacation. I spent time on vacation waiting to get back to work. Atari was *that* exciting, that rewarding, that much fun for me. Consequently, when it falls apart my grief is *that* excruciating.

As I carry "the box" through the parking lot on this chilly autumn evening, it strikes me I'm not coming back tomorrow. Atari, my dream-come-true, is shattered. I feel the shards crunching with every step. After more than a thousand coming-back-tomorrows, I've lost the only place I've ever really belonged.

It makes the chill of the evening air that much colder.

CHAPTER 18
TOUGH ACT TO FOLLOW

The intense heat of the high desert sun is punctuated by the occasional sand-blasting gust. The news crews do not appreciate this. Flying bits of garbage wreak havoc on sensitive equipment and delicate hairstyles. The entire site is littered with camera teams. Each one a separate little hive of activity, carefully positioned out of earshot of the others. This is a media circus.

Every crew features the Core Three: A camera operator, a lead producer (who calls the shots, literally) and the on-camera talent (easily identified by hair and clothing). I'm a checkbox on most of their shot sheets so I'm rotating between crews throughout the day. I'll do roughly fifteen interviews before that hot sun sets.

Sometimes events shape the media and sometimes the media shapes events, but right now one media team is shaping me, or trying to. The sun-bleached blonde is pretty, at least I imagine she is underneath the layers of blush and foundation. She takes my arm in a gesture of personal contact. Her smile suggests she's warming me up for the interview, but her grip makes it clear she's positioning me for the camera. I get it, she wants me facing the sun for better exposure during the shot. Good call on her part.

The only problem is I'm also facing the wind. We're in a lull now, but the next sand-laced blast can't be far away. I suspect we both know this. Feigning ignorance, I suggest we switch places, so the sun is not in my eyes. She appeals to my vanity, explaining how this will make me look better on camera. I resist and she insists. I turn one last time, my back to the sun and camera (and wind), and I make it clear that if she wants the interview, they should probably stand in front of me. She looks to the producer, they exchange nods and then reset to my position. Good call on my part.

Shortly after we get started, a vicious gust comes whipping through. My back takes the brunt of it gracefully, but this poor reporter gets an

unfortunate face full. It's not the sand in her nostrils, mouth and eyes that inspires my sympathy, it's the way her makeup holds the sand like glue. This gives her elaborately prepped face a frostbitten look, an odd counterpoint to the desert heat. I feel bad for her, but I'm glad it's not me (and I'm not even wearing lipstick).

The silver lining: After all this time, it looks like I've finally learned to spot trouble coming… and duck.

LIFE AFTER ATARI

Alas, poor Atari. I knew ye well. I had begged, pushed and stumbled my way onto a path that fulfilled my promise to my 15-year-old self. Now the path is gone, but the obligation remains. I'll need to find my next source of real satisfaction, if only I knew how.

Working at Atari meant never having to explain what you do. People knew Atari, and they were excited about it. Atari wasn't just a job, it was an identity. A self-explanatory self-image that made me interesting to people. It also made me interesting to me.

Some say losing the perfect job is worse than never finding it. I disagree. If lightning strikes once, I believe it can always strike again. That bolsters me. But the current forecast holds nary a storm cloud for the foreseeable future. This saddens me.

I should take some time off, just like the video game industry. I've got solid vacation money, but nothing approaching retirement funds. I decide to invest in an extended break, spending my days reading, watching, listening, playing and reconstituting.

Weeks turn into months, and months become a year. I feel the pressure building.

The revival of the video game industry will stand on Atari's shoulders, but that's easy to do. Once the world finished pile-driving Atari into the cement, their shoulders are at sidewalk level.

But where will I stand? What shall I do? Those pesky questions return. My inner 15-year-old reminds me of our vow, and he's calling in my marker. It's time to renew my quest to create a life worth living. But this time, as I take up the topic of next steps, I feel a pang of fear in my heart... and a pain in my ass.

A BRIEF COMPRESSION OF TIME

Though my E.T. game is already known as a bad game, it will still take more than a decade to become the worst game of all time. Mainly because a distinction of this magnitude requires two things: the internet, and an all-time.

In the meantime I need to do something, so I start doing everything.

My first stop was real estate. I want to try something completely different and, for a variety of reasons, real estate seems to suit my short- and long-term needs of the moment.

I earn a California Real Estate Salesperson license and start practicing commercial real estate in a small office. Don't care for it too much. Then I quickly upgrade to a Broker license and start practicing in a huge office. It turns out I hate working in real estate. It's no place for a computer programmer in Silicon Valley.

I switch back to tech, joining a mid-size company that makes systems for an enormous company. The work does not inspire me, but I meet many great people who do. One is a gentleman who wrote a book about ultimate frisbee. Each time I see it sitting there on his shelf I think: I want to write a book of my own someday.

Another more immediate inspiration comes from a fellow programmer who is going to law school at night. He shares his experience, which fascinates me. I'm interested in law and I need something more engaging, so I take the LSATs and score in the top 10%. That should get me in somewhere.

Just as I'm about to apply to law school, a headhunter presents an interesting opportunity. There's a startup in the field of high precision manufacturing. They work with complex design systems and industrial robots. That sounds cool. They also offer founders stock, the Silicon Valley siren call. Goodbye law school.

I jump into the startup fray, which means long hours, lots of tension and strain. But in return, we're going to get rich. Excellent!

Two years later the bubble pops and it seems clear this is not the winning ticket. This was an exciting ride, but it's over and I need to continue my journey.

I begin wandering around the tech industry, but technology is losing its considerable power to charm me. At this point, life bifurcates. Atari taught me that I'm not happy unless both my technical and artistic sides are engaged. I can't find this in one job, so I start working two: A tech job to feed my body, and a creative project to nourish my soul. My hope is some creative project will become substantial enough to sustain me and I can drop the tech jobs.

For my first project, I decide to write a book. I enjoy gambling, so I'll write a book about a gambling card game called Pan. But this is just a practice book to see what book writing is about. Then my second book, the book I really want to write, will be the better for it. And I believe Conquering College is indeed a better book. I even take some time off to promote it and hopefully make a splash. It does well enough to keep me working in engineering.

Like many people, I'm better at setting goals than achieving them. Though I enjoy spending time promoting my book, my money supply is dwindling. I will continue pursuing creative projects; exploring photography, writing magazine columns, giving seminars, etc. But right now I cannot afford to maintain my unemployed lifestyle so it's back to the cubicle for me. The double life continues.

ONCE UPON ATARI

I'm always exploring ideas for my next creative project, even if I'm not done with the current one. Often, coworkers are the source of this inspiration. While working at an industrial robotics company, my cubemate Paul informs me of a video production curriculum being offered at a local University of California campus. He knows of my interest in media and television. In addition to my engineering work, I would make amusing videos for company parties.

I look it over and decide getting a credential in video production will likely come in handy, so I go for it. Long story short, I complete their eighteen-month program in nine months and even get my final project aired on PBS (a first for the school). After creating a few promotional materials for the marketing department, I feel ready to begin a project I've been kicking around for some time.

Over the years, I'd seen many books and articles about Atari. The accounts were wildly inaccurate, largely because they were never told by Atari people. This bugs me, and the worst part of it is: Although they exaggerate for impact, their stories and events aren't nearly as sensational as the actual goings-on at Atari.

It also offends me on some level to see people misinforming the public about something so historically significant. When the truth is so amazing, why misrepresent it? But my offense over this feels like bullshit. On a more honest level, Atari was the most intense time of my life and I'm obviously not done with it yet. So, I decide to do something about it.

I became a video producer to better tell my Atari story… to myself.

Admittedly, my experience needed time to simmer. But after twelve years, it was time to garnish and serve it up.

So, in 1996 I begin a project that won't complete until 2003. I reconnect with friends and colleagues to produce the only piece of media created entirely by people who made games at Atari.

It has to be me. They are comfortable with me. I know the right questions to ask. And they can trust me because I'm just as guilty as they are, if not more so.

It takes me seven years to complete a documentary series which speaks the truth about the day-to-day experience of being a game maker at Atari. I do it to inform, but I also do it to heal. I need to make sense of my Atari experience before losing touch with it.

One of the toughest parts was coming up with a title. I rejected so many ideas. Ultimately, the winner appeared in a eureka moment and I recognized it instantly. I ran it by Jerome and he emphatically agreed. It had to be called: "Once Upon ATARI."

I shot it with a digital video camera, paying big bucks for the privilege. Over those years I spent a lot of time and money improving my video tech and occasionally popping out an episode. By 2003 I was finally able to put four episodes on a DVD, thereby completing the Once Upon ATARI documentary series.

I haven't talked with Steven Spielberg since he left the building after approving the E.T. video game. During the closing credits in each episode, "Hey Steven, Call Me!" falls quickly through the screen. He hasn't called yet. I remember my time with Steven quite vividly. I wonder if he remembers me.

YOU CAN NEVER GO HOME AGAIN

> "Play a game and you can waste a weekend.
> Learn to make games and you can
> waste an entire career."
>
> ANONYMOUS FISHING INSTRUCTOR

There is nothing like a brand-new millennium to make me reassess my life. Having stepped out of the game industry, I find myself out of step with my computational brethren (and sistren). I miss the exciting idiosyncratic nature of Atari. I also miss addressing both my creative and digestive needs with only one job. By 1999, I must face the fact that I need another dose. It's time to return to video games.

As I start looking around, it becomes clear to me things have changed in 15 years. There are lots more companies, and I wonder what the workplace is like now. It strikes me how vast the industry has become, and how narrow.

During one interview I find myself sitting with the head of game development. We're chatting enjoyably about the company's direction and plans to get there. I'll never forget it...

"Howard, our strategy is pretty straightforward. We are dedicated to continuing our current line of sports titles. That's what we do. It brings us a 10-15% return each year, which is what we want." Then he leans in for effect, "And what we're looking for is the big hit!"

I suppress the impulse to burst out laughing. I'm thinking this guy has a great sense of humor, he could be fun to work with. I'm waiting for the smile, or some indication he's joking...but nothing.

I brace myself and ask: "So if I understand you correctly, you're saying you want to keep doing exactly what you've been doing and you're looking for a *breakout* hit?" (pun intended, but lost)

"Exactly. Now tell me what part you want to play in this?"

I don't physically leave just then, but his Einsteinian absurdity assures me the interview is over. Clearly video game development has come a long way since Atari, but I may not care for its sense of direction.

HELLO 3DO

Eventually I reach out to my Atari network. It turns out many ex-Atarians are working at a company called 3DO. This is a company started by Trip Hawkins after he left Electronic Arts, which he also started. Trip is a major player in the history of video games, and an impressive character as well. They make me an offer and I accept.

[NOTE to the Name Origin Inquisitor: You may be wondering where the name 3DO comes from? According to company lore, it's the next big thing. There was radio, then video, now there's 3DO.]

I'm reunited with Tod & Rob, Bob Smith and a host of video game luminaries. There are also outstanding new contributors looking to make their mark. I believe 3DO is the most impressive collection of video game developer talent ever assembled under one roof. In retrospect, it's a pity we didn't get more from it.

I'm very excited to reconnect with the crew and resume my video game adventure. Nolan's Atari opened my eyes to the possibility of genuine job satisfaction, and Trip's 3DO offers the promise of regaining it. As I start to settle in, I'm fascinated by the similarities and differences from yesteryear.

One similarity is: You can always find someone working on their game. Day or night, weekends and holidays, just like Atari. But it's not really the same thing. At Atari this was a voluntary dysfunction, at 3DO it's mandatory. This is made clear in the offer letter.

3DO's offer letter states the company can cancel vacation plans if your project enters "crunch mode." They will refund money for plane tickets, but nothing else. Think about this… the company you might really need a break from can force you to stay there, and largely at your expense. That's what they're telling you up front. Imagine the things they don't want to put in the letter.

Like so many others, I signed it gladly because this is all about the dream: Producing an innovative piece of interactive entertainment, something groundbreaking that amuses and captivates millions of people while making you rich. That's the allure of video games. In this way, 3DO and Atari are very much alike.

But there's one thing 3DO has that Atari never could: There are people working here who grew up with video games. Whereas I longed to be an astronaut or a cowboy, they dreamed of becoming a video game designer/ programmer. Now they are pursuing their imagined ideal of a job, while

I'm picking up after the ellipses I left at the end of my time at Atari.

Whether building an imagined future or returning to a lost past, it's all about dreams. Dreams may be mini delusions, but they're the kind that make you smile, the ones worth chasing.

I also found similarity in Nolan and Trip. They both run companies and people; sometimes to the heights of success and sometimes into the ground. They both play hero and villain, but they do it truly believing things are going to work. It may be delusional but it's the kind of fantasy people want to buy into, and followers frequently wind up with a ride that's worth the sudden jolt at the end. How do they get people to follow them to the ends of their bank accounts and beyond? That's the power of charisma. Nolan and Trip are exciting compelling people who can sell the seductive notion of realizing dreams. Many talented people join them for this journey. And whether the final destination is riches or unemployment, most remain seated throughout the ride. You believe success is assured by joining them. That's the aura they project, and it is strong medicine.

Besides, when you're in a delusional place and you buy into the delusion, it can be a wonderful experience and remarkable things can happen.

At 3DO, I realize game development hasn't changed that much in some ways, people still spend tremendous amounts of time and energy doing the games, but now it feels more like obligation than inspiration. People still get stoned at work, but by the time I reached 3DO I wouldn't. Work is no longer cool enough to get stoned. Or maybe I've changed more than the environment.

Perhaps the most profound difference this time around is that making video games is no longer a work of authorship. At Atari, a game was all yours, the success was all yours and the failure was all yours. I really liked it that way.

Now it's a huge team in a giant collaborative effort. There's nothing wrong with collaborative development, but the sense of ownership is extremely diluted. This makes it hard to find the excitement and satisfaction I came here for. My best shot at returning to supreme job satisfaction has missed the target, and that's sad.

The industry has matured, the magic is gone, and I just don't care anymore... except for one thing: I do love the people. It's a tremendous collection of eccentric characters, and it's still a very dysfunctional and delusional environment, which has its charms. Yet another fabulous training ground on the road to becoming a therapist, too bad I'm not reaping that benefit yet.

Then again, some things never change. And one of those things is Tod Frye's ability to liven up a meeting. Most Friday afternoons, 3DO has an all-hands meeting. This one is the last company meeting before Christmas, which means 3DO will give some little gift to all the employees. This time it's miniature Swiss-army knives. They are tiny, with just a couple of blades, but they are personalized. Each one has the employee's name printed on it.

So, the gifts are bestowed, and our CEO Trip Hawkins is talking about the gifts. He notes how special it is that we have our names on them. At this point, Tod yells out, "They have the names on them so when you find one in your back, you know where to return it." The crowd roars with laughter, largely from recognizing the truth of it. This is pure Tod, pointing out both the humor and the brutality of developing video games.

Most delusional enterprises eventually lose their balance and fall. 3DO is no different. In late spring of 2003, they suddenly close their doors, leaving resolution of outstanding balances to a court proceeding (which will take years and consume much of the value).

The continuation of my video game journey ends in a calamitous reality crash. My denial is pierced and I'm adrift once again.

CHAPTER 19
SWEETER THE 2ND TIME

YOU WANT TO BE WHAT?!?

In a rare free moment, I find I'm still struck by the absurdity of hanging out in a garbage dump in the desert. Though I must admit, there is a certain poetry to it.

After 3DO blew up I wanted out of tech, so I focused on filmmaking for a while. I finished my Once Upon ATARI series, made several short promotional pieces and an award-winning documentary. I enjoyed it but couldn't sustain it. Since my body was still addicted to food, I figured it was time to swim back upstream to tech.

But this time tech was not answering. After a ton of resumes and a host of phone interviews, nothing was happening. It seemed being away a few years had rendered me obsolete. A relic. Not by my standards of course, but by theirs. It was OK for me to be done with tech, but the idea that tech might be done with me? To be discarded? That was shattering. It makes sitting in this garbage dump feel uncomfortably on the nose.

Fortunately, it was also a call to action. My switch from technology to psychology was not a whim, it promised some real benefits. One comes from the fact that technology is constantly changing and it's a pain in the ass to keep up. Though individuals are capable of enormous change, humanity is pretty consistent. The toys evolve but the games remain the same. Keeping up with humanity will be less work and more fun for me.

But it's more than convenience, it's alchemy. Tech is a youth obsessed field, and I was increasingly perceived as rusting, aging out. In therapy, age reads as experience. My move toward therapy is an act of professional alchemy, I'm turning rust into gold!

Telling friends and colleagues about my plan to switch careers was interesting. I remember the confused looks and quizzical expressions.

Many people (both in and out of tech) are flummoxed by the idea. They ask: How could a programmer ever be a therapist? I tell them it doesn't feel like such leap to me. After all, programmers and therapists are all systems analysts. I'm simply moving on to a much more sophisticated hardware: The human brain.

For others, the question is: Why do it? When you're in Silicon Valley and skilled in tech, people have trouble understanding why you might consider doing anything else.

Even my dear friend Jerome couldn't see it. Jerome and I were close friends for decades. If he hadn't died, we'd still be friends. Jerome is also the current record holder for most-talked-with-person in my life. For a few years, our main topic was his anxiety and frustration working in a startup. He would call most nights and we'd talk about it for hours. So, when I shared my plan with him, his reply struck me as odd: "You, a therapist? I don't get it."

I referred him to the years we'd spent discussing his work issues.

"Do you remember those conversations?"

"Yes."

"How did you feel afterwards?"

"Much better."

"That's why I want to be a therapist. I want to do that for other people too, and this time get paid for it."

I don't know if that settled it for him, because you don't always get closure talking with Jerome. What you do get is his next question. I never forgot it: "When did you come up with all this?"

That was a good one. Honestly, I don't think I've ever really answered it. Clearly it was before this conversation. There was the point where I applied to schools, but that's not really it. Endings and beginnings, the search always interests me.

I'm staring at the desert sands, an endless source of silicon, trying to figure out when I first decided to leave computer tech and become a therapist. It turns out the answer wasn't where I expected to find it. I remember the exact moment…

In late 2006, I'm taking one of my typical walks through a neighborhood park when I notice something distinctly atypical. I can't seem to breathe. I'm gasping for air. I'm lightheaded and feeling faint so I find a bench and sit for a while. My hands are tingly and I'm sweating profusely even though it's a cool day. Despite all this, I can't help noticing an oddly colored pebble sitting on the sidewalk by my feet. It occurs to me this pebble is probably having an average day with no trouble breathing whatsoever. Ordinarily I'd feel compelled to kick it on to the grass, but right now all I can think is: Wouldn't it be nice to swap lives with this pebble. Surely I'd feel so much better. I bet the pressure on my chest would ease.

I'm having a panic attack. This isn't good. Particularly when combined with the fact I've been waking up and spontaneously weeping lately due to an acute but unidentifiable sadness. These are new experiences for me, and alarming ones.

I'm in the middle of my third, and most intense depression ever. It's been decades since the last one. I'm currently unemployed and can't seem to find a tech job to save my life. Time and money are running out. My life isn't working.

None of my usual solutions seem to be available now, which is likely the reason I'm depressed and panicky. I know I have to do something, but I don't know what. Actually, it's more accurate to say I know what I want to do but I can't conceive of doing it… and I won't even know *that* for a couple of months. I'll learn it when I have "the conversation" with a woman I have yet to meet.

But for now, I'm deep in the throes of a solid depression, with anxiety to boot. This is also likely to be my last great depression, because this is truly the dark before a brilliant dawn. Unfortunately, I don't know that yet either. Once again, wellbeing is sacrificed to ignorance. For now, all I know is I'm sitting here on this bench, still more human than pebble, trying to catch my breath. Everything looks very dark with no dawn in sight.

I'm nearing thirty years here in fabulous Silicon Valley. It's hard to say if I've enjoyed it or merely survived it, but it's been one hell of a ride and I wouldn't trade it for anything. That said, if I were given the opportunity to edit it a bit, the present moment is one I'd likely delete. I'm experiencing a variant of a common Silicon Valley malady. Are you familiar with the Valley?

Silicon Valley is a remarkable place, thanks to a phenomenon I call the "Intellectual Gold Rush" of the late 20th century. It started with the idea that anyone with a vision and the grit to realize it could come here, hook up with venture capital and become a billionaire (or at least a multi-millionaire). Word spread and the floodgates opened. Hackers, visionaries, ideologues and influencers from every corner of the world rushed here to be in the place to be.

This gives Silicon Valley the illusion of tremendous diversity. Just look around, you'll see people from all over the globe. But the truth is: It's one of the least diverse places on earth. It's a frenzy of gifted, aggressively motivated people converging on one small peninsula to seek their fortune, squeezing out everyone with less drive or means or potential. Some succeed profoundly and get lots of press. Many more crash and burn and disappear. But the vast majority simply keep doing well enough to preserve the hope of doing way better, perpetually chasing a dream just beyond their grasp.

Silicon Valley is where the world's best, brightest and most ambitious people come to be average.

The boulevard of broken dreams runs right through the heart of Silicon Valley, and rush hour never ends. It's exhausting, trying to walk down the street when you're knee-deep in dashed hopes and crushed expectations.

People sideline their lives trying to bust through to a huge success. They pour their heart and soul into a start-up. It's right there in front of them, they see it, they want it, but they can't quite reach it. The allure is so great, they try again, and again. Some win. Most don't. It's all-consuming and it's brutal.

Here's what's bothering me: I don't want to play anymore. Technology no longer excites me, and I really don't know what else to do. Despite my best efforts, I'm still addicted to eating food and sleeping in a bed. At this point

I see only one reliable path to get my fix: A job in hi-tech. Unfortunately, I really don't want to do it.

On the other hand, since I'm not throwing my life away in the sinkhole of wealth-potential, I have time to do things like get out and about and connect with new people. In due course, I begin spending time with an interesting woman, a skilled healer in her own right. One day, while chatting enjoyably, I share my predicament.

She asks me: "What do you want to do?"

"I need to find a job in tech, like I always do."

"No, no. Forget about *need*. If you could do anything at all, with no restrictions, what would you *want* to do?"

"Oh, I want to be a therapist."

There it is. No hesitation and no doubt. My response is as automatic as breathing (except during panic attacks, of course). This is the instant I realize innately I want to be a psychotherapist.

All my life, people tell me their problems and seek my input. Friends, acquaintances and strangers alike, they all feel comfortable sharing their lives with me, I'm just that person. Usually they seem happier for it, and I enjoy it as well. I can't count the number of times people have told me I should be (or assumed I already was) a therapist.

I've had girlfriends say, "Now that we're dating, I don't need therapy anymore," which was awkward. Others have told me that I need therapy. I didn't disagree, but that was awkward too. Unsurprisingly, none of those relationships worked out.

When I was seventeen, I wanted to be a psychologist. Back in high school, my good friend Marc and I were planning to develop our own personality theory. Upon leaving high school we took radically different paths. Neither of us pursued psychology, but the desire never left. The proof? Thirty-seven years later, I became a licensed psychotherapist. Marc waited just sixteen years, then married one. He was always more efficient.

The healer continues, "Why don't you become a therapist?"

"What? I can't become a therapist. I don't have the time or the money for that. I've got to run my life; pay bills and meet expenses. My best shot at that is doing meaningless, soul-crushing work in an industry I'm no longer interested in or committed to."

After a brief giggle, she suggests, "Why don't you just look into it and find out what it takes to become a therapist? Maybe it's not as impossible as you think. Maybe you can at least start in that direction."

Programming may have lost its considerable power to charm me, but technology jobs haven't lost their considerable power to feed me. It's a quandary, a cognitive conspiracy keeping me from stepping into what I want (and perhaps need) to be.

On the other hand, why not? I've got nothing else cooking, and this seems a reasonable suggestion. In my previous job as a software manager, I focused more on addressing my engineers' personal issues than helping them with their tech. I discovered that most programmers don't need help getting motivated, they already love their work. They can, however, use help dealing with the issues that distract them from programming. I was already being a therapist at work (at least what I thought was a therapist) and I seemed to have a facility for it. But here's the big thing: I enjoyed it far more than the rest of the job. Dare I do what I like to do? Can I ignore all the practical considerations here? The last time I did anything like that was… going to Atari. Hmm. I decide to check it out.

Here's what I learned: To become a Licensed Marriage and Family Therapist in California (LMFT), I simply need to pass a couple of exams. No problem. However, before I can take these exams, I must have a master's degree (from a board-approved program) and complete three *thousand* hours of supervised training and experience. Problem.

I have a master's degree already, but it's the wrong one. I'll need another. And three thousand hours? That's a lot of hours. I'm turning fifty in five months. I mention to the healer how it feels a little late in my life to begin this kind of journey. As healers sometimes do, she tells me an interesting story about her parents… It seems her father was thinking of starting college at the ripe old age of 24. He was working, so he'd have to go at night. After figuring out the timing, he told his wife (her mom), "If I do this, I'll be 29 years old before I graduate!" His wife replied, "How old will you be if you don't do it?" He enrolled. Smart parents. Smart healer.

Once I start my investigation, it's stunning how smoothly things fall into place. It calls to mind another significant healer and mentor in my life, Dr. Gayle Pierce. She would frequently tell me that if I take one step toward life, life will take two steps toward me.

My great depression has been going on long enough and I can really use a win. Even if I can't get back into tech, one thing I can definitely do is get into school. So, I'll apply to graduate school and get some acceptance in my life. I identify one school, John F. Kennedy University, which really suits my schedule and feels good to me, but they're expensive. Then I notice Santa Clara University. They are less flexible in schedule and less simpatico in vibe, but they are more prestigious and more economical. It's also a school where I've been guest lecturing, so I know and enjoy several of the faculty members. I'll apply there. No interviews are required, and with my previous academic record this should be a breeze.

After applying, a question occurs to me: How will I pay for the whole thing? Money is tight already, tech isn't calling me back, so maybe I can start doing something else while pursuing essential credentials. I decide to consult a career counselor for advice on options. After interviewing me for an hour and a half, he offers me a job working for him, and I take it. That was easy. It doesn't pay a lot, but I'm already working in the field. How cool is this? Then, within a month, one of those elusive tech jobs suddenly appears… and does pay a lot. Bingo! I wonder how that pebble is doing these days.

Things just keep falling into place. So much so, I can barely take full credit for becoming a therapist. The universe is stepping in to smooth the path significantly. Of course, it was not entirely without bumps. While one universe welcomes me, another one rejects me.

I receive a letter stating that Santa Clara University has elected to NOT accept me for admission. This cuts me deeply. It reverses the recent uptick and further damages my already beleaguered confidence. Score one for the darkness.

But I don't try to beg my way in like I did with Atari. Instead, I challenge their decision. My quest to discover what went wrong winds up revealing how my tenure there would have been a nightmare. Good job, universe! The rejection stings, but once again I trip and fall down only to wind up dodging a bullet.

I refocus on John F. Kennedy University and enroll a few weeks later. As I originally suspected, JFK *is* the right place for me. A bit more money, but it's the program I need.

It's all coming together. I'm going to school, the job is covering my expenses, life is indeed taking two steps toward me. Impossible as it seemed to find anything in tech, once I earnestly pursued becoming a therapist, the right tech job found me. Two years later, I'm approaching graduation and I need more time to concentrate on accumulating my 3,000 hours. As I'm trying to figure out how to balance the job and the internships, I suddenly get laid off. This not only solves my problem, but also puts me in the running for the coveted Most-Timely-Job-Termination award.

Everything in my life is lining up for me to become a therapist... except me. I have some work to do myself, beyond just taking the tests and getting the hours. It's a good thing that job came along. In addition to funding my schooling, it also covers my personal therapy.

You may have noticed that much of my life to this point has been about doing things. I was never much of a be-er. If I'm going to be a therapist, this has to change.

My style needs a makeover, too. It's always easier to be the exception than it is to be exceptional. Though I believe I'm capable of both, I'm lazy and too often settle for being the exception. This time, I'm ignoring my inner cynic and making a real commitment to growth. Becoming a therapist is making me ask more of myself, and I'm sincerely trying to deliver.

WHAT'S SO GREAT ABOUT THERAPY?

I thought becoming a therapist would make me happy, but it turned out to be the single best decision of my entire professional life.

For me, therapy is a brilliant mixture of art and science, posing both creative and technical challenges. Atari taught me how I need both, but therapy is the only job I've found that combines the two more elegantly than Atari. And I looked in *a lot* of places. It was rough spending twenty-five years seeking and not finding. But on the upside, it didn't take thirty.

Another wonderful thing about therapy: It makes every bit of my background relevant. The programmer, the filmmaker, the economist, the writer, the teacher, the photographer, the real estate broker, the student, the director, the manager and the performer all provide value to my clients. My esoteric background helps me understand a wide variety of predicaments and perspectives. This allows my clients to spend less time explaining and more time relating and healing. I've struggled to find career opportunities that bring more of my experience into play. Practicing therapy does this in a beautiful way.

[NOTE to the Career Mired: One of the real impediments to changing careers is starting at the bottom again. The prospect of walking away from hard earned expertise and status can be enough to prevent anyone from changing paths at a time when maybe they should. Fear is powerful, but misery can loosen fear's grip. Changing careers was easier for me because I relish the steep part of the learning curve, but mostly it was intense dissatisfaction with where I was.]

And, on a personal level, I have been depressed, anxious and insecure. I've had plenty of issues and problems, and I learned to deal with them. I know what it's like to lack boundaries, and then to develop and hold them. I know what it's like to get divorced and what it's like to create a healthy relationship. I've had dramatic successes and huge failures. It wasn't an easy road, going through such a litany of circumstances in life, career and wellbeing, but I'm grateful for the trip. It helps me help others to understand where they are and plan their next steps more confidently.

I'm tremendously passionate about therapy. It took three decades to feel that again, but it was worth the wait. I'm doing things of genuine import, impacting people's lives in a positive way. Psychotherapy is challenging, intimate, draining, rewarding, even scary at times. It is by far the most deeply meaningful work I've ever done.

That's what's so great about therapy.

THE PHENOMENON OF WORST GAME EVER

E.T. is frequently cited as the worst video game of all time. It was clear from the start there were issues with the game, and it crushed a lot of Christmas dreams in 1982. Some people have even suggested I became a psychotherapist to help address all the trauma and depression created by releasing it on an unsuspecting public. But worst video game of all time? That's quite an accomplishment. I promise you that was not my goal.

In fact, when the E.T. game was done, that wasn't even a possibility. It was the early '80s. There was no internet, no downloading and no instant player feedback. It was the dawning of video games. There were no oldies, only newies. You can't have a worst game of all-time until there's an "all-time".

Through the '80s, E.T. was not the focus of why the industry crashed. The focus was mainly on the mere fact that it crashed, thus proving that video games were just a fad and not a serious prospect. This opinion circulated among many investors, which makes a huge difference in the perception of lots of other people.

The advent of E.T. as the face of the crash didn't get going until well into the '90s. Things really started to happen as the internet came on, with its insatiable hunger for content. It turns out lists are a great source of net content. Everywhere you look there's Top-5 this and 10-Worst that. Video games fit the bill perfectly. Often I'd see Yars' Revenge gracing the Top lists and ET anchoring the Worst.

It took more than a decade for "all-time" to reach critical mass. But once there, the internet assured us my E.T. game was responsible for perpetrating unspeakable acts upon the fledgling industry. The phenomenon of Worst Game of All Time was born.

Then came hater culture. The E.T. game got even more attention when it became a lightning rod for haters. I never argue with player opinions. If someone says they don't like something I have no reason to doubt them. However, when people make a point of telling me how E.T. is such a horrible game, I like to ask: Have you ever played it? It's amazing how frequently the answer is no.

Once the calendar struck 2000, classic gaming conventions were popping up everywhere, and growing. I'd be invited to appear on panels and give talks. It was fun to meet players and reminisce about Atari times. It was fun to reunite with old friends and colleagues. I also started writing columns for video game magazines.

Over time, the urban myth of E.T. games buried in the desert gathered steam. There were articles, posts, songs, videos and all manner of speculation about the existence of buried treasure, and the prospect of a map. People opined about it, sham "discoveries" made the social media rounds, but the interest persisted.

Ultimately all this attention and speculation resulted in a budget and a film crew. That's why I'm in New Mexico, staring at mountains to the east, an endless supply of sand to the west, and a food truck right in front of me. I hope it's not a mirage.

CHAPTER 20
GROUND BREAKING

The big noisy yellow monsters are working away. Undaunted by the sweltering afternoon sun, nothing stops them from delivering scoop after scoop of ancient trash. Arranged in neat mounds across the desert floor, the area inside the plastic retaining fence looks like a giant braille message. And so far, it does not read: HERE THEY ARE!

Surveying this massive collection of garbage, we're waiting to see if something big will happen. It's just like Atari, anticipation with no sure payoff. It's a reliable formula… for tension.

I've always been sure there wasn't anything here. However, I do remember one day when I said the cartridges were there. I had to. It was in the script. It was almost exactly two years ago…

WHAT IS A VIDEO GAME NERD AND WHY IS HE ANGRY?

It all started with another fateful call, this one in the form of an email. In 2011, I was approached by CineMassacre Films with a script they had with an entirely new take on the whole E.T. story. The Angry Video Game Nerd (AVGN) was doing a movie. James Rolfe is the creator and star of AVGN. He is a devoted filmmaker who invented a character called the Angry Video Game Nerd. The Nerd, as he is known, stars in a series of web videos in which he reviews the worst games from the old consoles. He plays the games, all the while spewing incredibly colorful and amusing insults as he suffers through the experience. Invariably, the pain of playing the game becomes so great, the Nerd goes apoplectic and destroys the game cartridge in some outrageous over-the-top flourish. It's quite entertaining. He developed a substantial following over the course of the series. Since his focus is the worst classic video games, his fan base wanted him to review E.T. For years they begged him, and he steadfastly refused. Is E.T. so bad, even the Nerd won't touch it?

It turns out this was part of a long-term plan. After years of building a substantial web following, he opened an IndieGoGo campaign to fund an AVGN movie. They raised over $325,000! Pretty impressive for a Nerd.

I read the script and had some objections about my character, so I did something virtually unprecedented in the history of film acting, I asked for a smaller part. And I got it. I suggested they add a level of deception, inventing another character who acted through me instead of me acting directly. They liked it and rewrote the story. I blessed it and away they went.

Wrote a column about it. Like to read it? With the permission of Imagine Publishing, here it is:

"Ha! You'll never live *that* down." This is a phrase usually associated with pub crawls, late night drunk-texting, or the best-intentioned public moments gone horribly awry. We all have one or two. Remarkable nights of debauchery, stunning flashes of stupidity and the occasional confusion between thinking and saying. That's the stuff of which Never-Live-That-Down moments are made. Oh yes, and there is one other category that appears to fit the bill: video games.

The funny thing about Never-Live-That-Down moments is how they evaporate into the ether quicker than we ever thought possible. Not mine though. I made the E.T. video game for the Atari VCS. Whatever one might say about that game and the dubious circumstances surrounding it (most of which has already been said multiple times in 30 years), one salient fact remains crystal clear to me: I will never live it down.

How do I know this? Simple. They sent me the script.

Over the years there have been many allusions to and discussions of the great Alamogordo, New Mexico dump site and its alleged substrata of E.T. game cartridges. There have been reports, songs, investigations, articles, YouTube videos, infrared analysis, folk lore/mythology and literally hundreds of interviews on this subject. The New York Times, Snopes, Wall

Street Journal, Wikipedia, IGN and uncountable forums and blogs opine about it. All this activity principally comes down to "Is it true? Are there millions of your game cartridges buried in the desert?" I've been hearing, speaking, reading, seeing and joking about this for nearly three decades. But just when you think there couldn't possibly be anything new to say about the E.T. fiasco, someone comes along with something new to say about it.

This time, that someone is James Rolfe of CineMassacre Productions (creators of the *"Angry Video Game Nerd"* video series). He and his cohorts are putting together a movie about the E.T.-carts-buried-in-the-desert controversy, and as I said before, they sent me the script.

Why did they send me the script? Well, it turns out I'm actually IN the movie. So they figured since I'm a real person and I'm in the movie and I'm actually me and they wanted to use me as me in the actual movie and I didn't even know yet that I was me in the movie although I did know I was me but I didn't know I was in the movie, they had better let me know that I was in the movie as me and let me see if after I knew I was me playing me in the movie that I would be OK with being me in the movie as myself now that I knew there was actually a movie with me in it. Maybe that's too much legalese, but that's what movies are all about.

I read the script. I'm not issuing any spoiler alerts here because I'm not issuing any spoilers, but I do have to say one thing: This is the freshest take I've heard on the subject in 30 years, and I've heard a LOT of takes on this subject. After all these years someone surprises me with a script which redefines my relationship with E.T. Astounding!

This reassures me I will never live E.T. down. I don't regret it. I enjoy its infamy. I still get a kick knowing I made the worst game of all time and it feels cool to be powerful enough to have brought a billion-dollar industry to its knees with just 8K of 6502 assembly code. People frequently ask me: How can you be so happy knowing you made the worst game of all time? I tell them it's merely pragmatism. I'll never live it down so I might as well live it up.

[NOTE to the Referential Reader: This column was written in December of 2011. It appeared in issue #116 of GamesTM magazine in the United Kingdom. It is reprinted here by permission granted from the good people at Imagine Publishing.]

A few months later, in the spring of 2012, I flew down to southern California and rode out into the desert where the filming was taking place. I was impressed. It was a very professional shoot. I didn't study the part very long because I had been living it for so many years. We got all set up, and when the time came for my close up, I gave the line: "The legend of the buried carts is TRUE!" This was the first and only time I declared that game cartridges are actually buried in the desert. I was pointing to a different desert and I was just reading lines from a script, but I said it with conviction and it made the final cut of the Angry Video Game Nerd Movie.

Now I'm here in the actual location waiting to find out if my fictional statement will blossom into a full-fledged prognostication.

WHAT'S ALL THE COMMOTION?

April 26th, 2014 is not only the day of the Alamogordo dig, it's also my mother's 78th birthday. How perfect is that? Without her, I wouldn't be here. Of course, with her I might not be here either. She didn't want me to go to Atari. When I announced I was leaving Hewlett-Packard to go make games, she told me I was throwing my life away. She told me I wasn't her son, because no child of hers would do such a stupid thing. She came around though. After I made several million-sellers and put an addition on her home, she told me it was a good thing I had listened to her and gone into computers. This may shed some light on how my background prepared me for becoming a therapist, and before that a client. After all, if it weren't for families, there would be no therapists.

But I *am* here. I'm signing autographs, doing interviews and having a wonderful time with lots of interesting, talented and delightful people. And all because three decades ago I did something that touched their lives.

Looks like I'm a karmic lottery winner after all!

I'm also hungry. Sherri, my fabulous wife and travel companion, is hungry too. Our blood sugar is starting to crash after six hours of all this. We agreed she should get out of the sun and go relax in the trailer while I get some food and join her. I'm waiting for my order at a food truck when the hoard starts to swarm. A huge people-mass is now pressing on the fence around the excavation site. I finally receive our meals and start heading for the trailer when two production assistants rush up to me in a panic, "Come on, we gotta go!"

"I need to get this food to Sherri first." One of the assistants grabs the food from my hands while the other gets behind me and starts pushing, hard! Before I can say "excuse me," the crowd is behind me and I'm squeezed against the fence. I'm facing the garbage-pile-studded excavation area. Staring back at me from inside the fence is Zak Penn, hard-hatted film director. He is standing there with a microphone in one hand and a bucket in the other. He moves the mic close to his mouth, "We found something…"

WE FOUND IT!

There is triumph in his voice and relief on his face. Zak reaches into the bucket and pulls out a partially crushed but totally discernible cardboard package. "…E.T. the video game. Intact. In its box."

A chorus of cheers and shutter clicks rise from the crowd. Zak's heavily gloved hands fumble with the cardboard as he retrieves the contents, a chipped and flattened E.T. video game cartridge. Zak hands it to me for authentication.

I've spent many years distancing myself from this story for my own protection, but there is no distance now. This ancient piece of silicon and plastic is sitting right here in my hand. I can feel the rough texture of its cracked exterior grating against my bare palm. I realize he is wearing protective equipment from head to toe, and I have a brief Marie Curie moment. Thanks for the microbes, Zak.

Then I verify the cartridge and hand it back to Zak. He holds it aloft and the crowd goes wild! People are cheering and shouting, high-fiving and hugging. The air is filled with excitement and wonder (and dust)! Suddenly, what seems like a hundred cameras and boom mics are in my face, "Hey Howard, what are you feeling now?"

My impulse is to perform, but an incredible rush of emotion pours over me. Everything goes eerily quiet. I feel things welling up inside. For me, the importance of making games is to entertain and amuse people. To give them a break from day-to-day life by creating wonderful moments. And now, in the middle of the New Mexico desert, a piece of work I did 32 years ago is doing exactly that for hundreds of people. As I drink in the excitement my game is generating, my heart fills my throat and tears pool in my eyes.

This thing I created long ago with an excruciating effort, followed by an extended history of strife and criticism… it lives! I try to speak but joy and gratitude overwhelm me.

IT TURNS OUT YOU CAN GO HOME

After a weekend of literally digging up my past, I'm sitting in the El Paso airport waiting for the flight home. I just got off the phone from wishing my mother a belated happy birthday. I was so wrapped up in everything around the dig I forgot about that entirely.

I got to revisit the remnants of an old life journey, just as I'm hitting my stride in a new one. Endings and beginnings. After Atari, my professional life was a wasteland for many years and many jobs. They all had their moments, but psychotherapy is my true calling. I always believed I'd get there someday, but decades wandering in the career desert really tested my faith.

My phone rings and it's Barn, my college roommate and loooongtime friend. He's calling from the Florida hospital where he's facing the end of a decade long bout with multiple myeloma.

I flew out to be with him a few weeks ago, and we've chatted a few times since. But with each of us in our respective boarding areas, it's clear we are speaking for the last time. The call is brief, a discussion we've never had before in our thirty-nine-year relationship. I tell him I love him. It turns out he loves me too.

HOW DO I FEEL ABOUT THE WHOLE E.T. THING?

As impossible as I thought it to be, the games were there. It made no sense, but *when you expect things to make sense, you're losing touch with Atari.*

It was a glorious affirmation of just how crazy Atari was, and that's what made Atari so incredible. It was a hotbed of abject excess that could never last and could never be replaced. The Atari I knew and loved evaporated in 1984, and soon thereafter I left. But where do you go after an experience like that? Apparently, you wind up in a sandstorm in a dump in the desert.

Much of my experience has been people expecting me to feel bad or be hypersensitive about E.T. Doesn't it bother me to have made the worst video game of all time? How does it feel to have spawned an industry crushing failure? They don't seem to get how I can have a positive take on the whole thing. Honestly, E.T. never quite felt like a failure to me and I'm happy to explain why. (Hint: it's not because I'm in total denial) (at least not this time)

Failure holds a noble place in my mind. It provides an opportunity to find the cracks in my worldview and fix them. Failure can be good stuff if I use it wisely.

On the other hand, failing hurts. I may value it, but I don't enjoy it. So, how do I feel about having made one of the worst failures of my generation? Well, overall…

I feel great about it! I accomplished something most people thought was impossible. I was able to meet an aggressive personal challenge and deliver one of the highest selling games for the Atari VCS in the shortest time ever, by far.

E.T. completed the range of my work. One of my games is in the New York Museum of Modern Art, and another is the subflooring of the New Mexico desert. I've always preferred breadth to depth. As a game designer, E.T. showed me I can make a game for anything.

Also, this game I produced over 30 years ago is still a media topic. How many game makers can say that?

I've had many successes in my life, but I'm best known for one of my failures. One of my successes is not letting this fact make me feel like a failure.

Every failure offers a choice; focus on the shame or the lessons. I tried shame for a while but settled on lessons. I learned that surviving failure takes resiliency, and resiliency is best cultivated by pondering the right questions.

Was the E.T. video game a failure?

I guess it depends on who you ask. Salespeople say that even after subtracting returns it sold over 1.5 million units. That's good. But accountants note that Atari bought 4 million units and paid a huge amount for the license so the product lost money. That's bad.

If you ask me, it sure felt like a success for the first few months. It was completed on time, then the sales feedback was excellent. Remember, there was no internet then. After a while, it became clear people were not liking it and the returns overwhelmed the picture. Failure is tough, but when an initial success does a U-turn to become a failure, that's tougher. After years of receiving both criticism and fan mail about the game, the question of success vs. failure feels more like an ongoing struggle.

But if you could ask Jerome today, he would be far more definitive on the topic. Jerome was all about media. And one of his favorite things in media is references to other media. Most people would appreciate a few extra years of life, but in Jerome's case it would have blown his mind. Because in 2019, our E.T. game was featured in the Simpsons' Treehouse of Horror episode. Jerome and I were into the Simpsons. WAY into the Simpsons. Especially the earlier, funnier seasons. For instance...

When Sherri and I honeymooned in Paris, we were jet-unlagging in the room and a Simpsons episode came on TV, in French. I began translating the dialogue for her line by line. She was impressed.

"How do you know what they're saying?"

"I studied it for years."

"You studied French?"

"No. I studied the Simpsons."

Through all that schooling, Jerome was my classmate. Seeing our work immortalized in a Simpsons episode would constitute irrefutable success in his eyes. It didn't do me any harm either.

Do I believe E.T. is the worst game of all time?

Not for a second. In fact, many people love the game. But do I prefer when people call it that? Absolutely! I also did Yars' Revenge, which is frequently cited as one of the best games of all time. So as long as E.T. is one of the worst, I have the greatest range of any game designer in history! (a sense of humor is essential to resiliency)

Having done one the best and worst games of all time, what's the difference?

I believe they are both innovative games with solid design concepts and good basic mechanics. The crucial difference between the two games is that one was done in seven months and the other was done in five weeks. Does this mean I assume, given enough time, I can reliably make a good video game? Yes, it does. I believe my track record bears this out.

E.T. was my third consecutive million-seller. Would this game have sold a million without the license?

No way! But then again, without the license there wouldn't have been a five-week deadline. I'd have time to grow it into a full-fledged HSW production. At two months Yars' Revenge was a horrible game, but an additional five months turned it into a groundbreaking, industry-standard setting success. Given more time for E.T., who knows what might have been?

Is Steven Spielberg an alien?

I like this theory, and I held on to it for a long time. Pro: He abducted me professionally and *that* close encounter changed my life forever. Con: When he directed a remake of "War of the Worlds", that pretty much killed the whole concept for me.

Jerome and I spent countless hours chatting enjoyably, and our fascination with media accounted for the lion's share of that time. Anyone who ever joined us for a movie knows this. The ensuing hours of analysis and criticism was proof positive.

Media production was foundational for both of us, so let's consider the question of success or failure from that perspective. I believe the ultimate goal of media is to entertain, to inform and to generate social discourse (or for the haters, DIS-course). Here we are, discussing it nearly four decades later. That feels like a win to me.

What's my take on making the worst video game of all time?

It's quite an accomplishment, but not necessarily a public service.

CHAPTER 21
IMPACT

Many products start as fads. Some make the leap to lifestyle innovation and become an indelible part of our day-to-day existence.

One of the remarkable things about working at Atari was the pervasive feeling we were changing the world. We could see ourselves revolutionizing the landscape of entertainment and the meaning of television, yet time would prove we grasped precious few of the myriad ways video games would ultimately shape our lives.

The games we made and the games we played at Atari were merely the microcosm. The impact of video games on society and the world, that's the big picture.

VIDEO GAMES CHANGED THE WORLD

Video games didn't just change how teenagers spend downtime. They grew to permeate every facet of our lives.

Atari's contributions, both successes and failures, became the roadmap everyone in tech development followed.

First and foremost, video games made television interactive. That's huge! Change television and you change the world. Thanks to video games you are no longer the zombie sitting on the couch passively staring at the screen. You are now the zombie with a controller actively yelling at the screen.

Interactivity is a big thing with me, and not just in games. My therapeutic style is highly interactive, using lots of back and forth with my clients. This is because therapy relies on the original interactive medium: Conversation.

Video games also turned out to be the killer app for home computing. Atari put millions of computers inside homes at a time when IBM was only crossing the 10,000 mark for home computers. Everyone was trying to answer the question: Why do I need a computer at home? Apparently, the correct answer was: To play fun games!

Video game technology changed the economics of simulation, which has revolutionized the fields of design, training, construction, transportation and entertainment, just to name a few.

Video games made virtual reality, augmented reality, environmental simulation (and many other hi-tech buzzwords) into regular features of our day-to-day life. It used to take huge government budgets and the Department of Defense to create new technologies. Now, thanks to tools developed for video game production, it can be accomplished by a few people in a garage.

And to paraphrase Shakespeare: Let's not forget the lawyers. Our legal system has evolved significantly due to video games in general, and Atari in particular. The industry created scenarios and test cases which led to precedent setting decisions in trademark, licensing, patent and intellectual property law. This redefined how people create technology and deliver content. Big stuff.

Video games changed the workforce and the workplace. In 2020, video games employ over 240,000 people, and the industry is still growing. These people make a living designing games, programming games, creating art for games and yes, some even make a living playing games (we call them testers).

In addition to the 240K game employees, more and more people are making a living by *being good* at playing games. Money making opportunities inside video games are also growing. Excellent players run around popular game environments collecting high value assets, then turn around and sell them off to aspiring champions with more money than time or skill. Secondary markets emerge, increasing the dollar value of the video gaming marketplace. Think about that: Pseudo-worlds are creating real life economies. That's a pretty amazing thing to have grown out of Pong!

Then there's the arena aspect. Video games are now a spectator sport. As with any activity, the people who pursue it tend to enjoy watching others who do it extremely well. The audience is growing, therefore the advertising possibilities are growing, therefore the prize money is growing, therefore the possibility to make a living as a professional video game competitor is growing. Can you beat that? If so, you might be able to quit your day job.

Are video games art? This is a great topic I'm not debating here. However, there is no debate that video games are their own medium. I believe any medium can potentially host art, and I'm proud to say the New York Museum of Modern Art agrees. They proved this in 2013, by selecting Yars' Revenge as part of a modern art exhibit devoted to video games. This not only makes me a MoMA exhibiting artist, it removes a biggie from my bucket list!

Video games aren't all fun and games. As a new medium, it impacts other aspects of life as well. In some cases, video games can become an outlet for behavioral addictions. They join the likes of gambling, sex, eating and other activities in this regard. I deal with this regularly in my therapy practice and trainings.

This leads to another significant life task invaded by video games: Parenting. Caring parents spend too much time torturing themselves with: Are we doing this right? Video games have added a new and puzzling wrinkle to this age-old tradition. This is exacerbated by research, media and ever more seminars and trainings. When it comes to parenting, video games can make a tough job tougher.

Addiction and parenting are deep, serious and complex issues. Rest assured, lots of time, money and attention goes into addressing them every day.

Video games also do some wonderful things. Right from the start, they were used to rehabilitate stroke victims by facilitating redevelopment of eye-hand coordination. In fact, video games have therapeutic applications for a wide variety of issues. Over time, the gamification of certain day-to-day activities has proved an effective motivator for exercise, diet and training applications.

Video games give me a sense of agency. It's a world where my skills and abilities determine the outcome, as opposed to the whims of others (specifically bosses and parents). You can reliably accomplish things in games. That doesn't always happen in real life.

Video games can alleviate boredom and provide consistent fun. They can also capture my attention fully, offering me a break from other tensions which might be weighing on me.

Whereas some people just need a break, others need a full-on escape. People who feel stuck in abusive situations report that video games can provide temporary relief from acute feelings of powerlessness. That may not sound like much, but temporary relief in a desperate moment can save lives.

ATARI LEGACY

The great video game crash is a tale of hubris gone awry. Right place, chaotic time. It's about experts from a variety of fields joining together in a situation no one had ever seen before and expecting to make it work as if they'd been doing it all their lives. We bought into it because of BMOBS. We thoroughly, deeply and fundamentally believed our own bullshit.

Warner hired Ray Kassar to solve a problem they didn't understand because nobody knew anything. Maybe they should have hired William Goldman. But in retrospect, it's easy to see how Warner's top brass saw Ray as a sensible choice for the job.

That's the story of Atari… a bunch of smart people making sensible choices as we all join hands and walk off a cliff.

Yet one of the main ways Atari changed the world was by legacy. Everyone at Atari held three fundamental beliefs:

1. Things can take off at any moment.
2. Things can fall apart at any moment.
3. A single product can change the world.

Every Atarian carried this DNA to their next company. That's 10,000 seeds spreading Atari culture throughout Silicon Valley and the world. Each one looking to sprout the next round of economic lightning strikes. Some succeeded.

I wonder how much of the dot-com bubble (boom *and* bust) was the fruit of these 10,000 seeds? And how many people fell down a rabbit hole following an ex-Atari dreamer?

Atari helped forge the essential contract of Silicon Valley. This contract states that if you're willing to blow off your entire life for Success (your relationships, your family, your friends and yourself) you can get paid-off really well… maybe.

THE IRONIES OF E.T. & ATARI

For someone who loves irony as I do, Atari provided a generous feast.

First there are the ironies of the E.T. project itself. For instance, one of the most expensive movie licenses of all time was given the shortest schedule.

One of the most popular movies of all time yielded one of the worst games of all time

In order to complete the project, I had to cut myself off from my main source of inspiration, my coworkers.

The game that had trouble with pits wound up in a pit. Of course, this may be better filed under Poetic Justice.

The worst video game was the most successful media product. After all, we're still talking about it. This dovetails with the fact that the game I gave the least time garners the most attention.

Then there are the ironies of Atari in my life. Like the fact I avoided computers until the last possible minute, then made them my career. Well, first career, anyway.

My entry to Atari was ironic because it nearly wasn't. I wore a sport jacket to the interview and acted in a highly professional manor (i.e. artificially stiff). This is not what they were looking for at Atari. My effort to disguise myself to get the job almost locked me out of the perfect job for me.

One of the great ironic Atari moments in my life came many years after I left. During my studies to become a psychotherapist, there were occasional seminars and trainings on weekends which couldn't be missed. There were also Classic Gaming Conventions which were great opportunities for me to relax and enjoy my friends and community. I really valued those getaways. There was one time where a show came up on the same weekend as a psychopharmacology class I needed. It was an interesting anomaly that drugs kept me away from an Atari-related event.

It is ironic that making video games is such hard work. Creating solid entertainment is extremely difficult and taxing.

Then there's the irony of this book: I did not think it would take me years to do this. Along the way I kept setting arbitrary deadlines. How can I write about a ridiculously short schedule without a deadline? Fortunately, I blew them off each time. I'm happy with this result because I allowed myself the time needed to get here. It amuses me to think I might have diminished a book about the E.T. video game project by enforcing a shorter deadline.

But the biggest irony of all has got to be that an entirely non-violent game killed a multibillion-dollar industry.

THE SAN DIEGO SCREENING

Three months after Alamogordo, Sherri and I find ourselves in San Diego for Comic-Con. This is exciting because I've always wanted to go to San Diego Comic-Con. It's also nerve-racking because I'm going to the first screening of Atari: Game Over. Finally we will get to see the result of all the interviews and our day in Alamogordo. When I think of how intense and revealing those interviews were, I'm a little nervous about what might appear on the screen.

The dazzling array of costumes and inventions amazes us as we stroll around the con. I'm always impressed by the creativity and ingenuity fans bring to their beloved pursuits.

And today I'm not just an attendee, I'm a panelist. One of the advantages of which is being able to hang out behind the scenes, something I always enjoy. As we near the area where the movie production people are gathered, more passersby recognize us from the dig (and from the film footage, apparently).

They make passing comments, such as: "Oh, you're really going to like it," or "You're going to be so happy with this." This harkens back to the hallways of Atari with people telling me, "we don't blame you," and "it's not your fault." As I think of all that proceeded from that, it adds a touch of anxiety to my cinematic anticipation.

The evening arrives, as do we at the scene. It's a beautiful new building with lovely décor and brilliant appointments. An usher greets us, dispensing good cheer and popcorn, then beckons us to follow. This is the first time since the London premiere of E.T. that I'm escorted to a reserved seat in a movie theater. I turn around to see if Spielberg is three rows back. He isn't. But with Nolan Bushnell on my left and Sherri on my right, it feels like the best seat in the house.

Geometric center of the theater doesn't hurt either. The cushions are comfortable, and the popcorn is perfect. I'm well situated in a theater ready to watch a movie. I've done this thousands of times in my life. This time, however, is not the usual.

The film's director is here to introduce it, that's only happened a few times. And I'm *in* the movie, which has never happened before. I mean, it has in dreams and fantasies, but this moment is far more life-like. Eerily so. It's a very different feeling.

Zak stands up by the screen and says some things that my brain is way too preoccupied to register. I grab Sherri's hand for comfort. I double check to make sure it isn't Nolan's. And they roll the film…

There it is on the screen, bits and flashes of my journey. A story so calamitous, Hollywood had to make the movie. Complete with analysis and commentary from key players, industry luminaries and me. The whole thing feels... weird.

After a tumultuous sixty-six minutes, Atari: Game Over is over. My instant reaction: I really like it. More importantly, I am proud to be a part of it. But that's just the tip of my emotional iceberg.

Zak is back with the microphone. He's adding some interesting comments about the making of the film. At least I presume they are interesting because Zak is an interesting person, but I can't quite focus. This film has brought on a rush of thoughts and feelings.

I've gone over this story so many times in the last thirty years. I've given over a hundred interviews, discussed it on panels, in lectures and even the occasional keynote. I've laughed about it, cried about it, explored it thoroughly… I should be done with it already. Right?

Sherri gently nudges me out of my reverie. Zak is calling out audience members who are in the film. Zak introduces Joe Lewandowski, the person chiefly responsible for identifying the right place to dig, and everyone claps. Then Zak introduces Nolan Bushnell, the person chiefly responsible for Atari, and they clap more.

But if I'm so over it, why is this so intense for me?

Then Zak introduces me, the person chiefly responsible for the worst game of all time, and the crowd erupts into a standing ovation.

I stand up too, and look around. Every person in the theater is standing, staring at me and clapping. I'm flooded. Time dilates. I'm overwhelmed by an unfamiliar mixture of emotions. Attempting to make sense of it, my mind wanders...

I didn't go to Atari for the games, I went for the work and the environment. However, the games brought me so much more.

In my heart I was always an artist, but I was afraid to own it. Atari gave me the gift of opening my self-concept and broadening it.

Atari taught me no matter how dark life seems, it can turn around entirely in just two years! This knowledge inspires me.

Atari made me a therapist, eventually. When Zak handed me the cartridge at the dig, the intense joy I felt was only accessible to me because of my work in becoming a therapist. Atari was the best four years of my first 50. But after that things really picked up.

Atari showed me where I belong. It helped me find my path to satisfaction in life, and eventually to spot the pits...

I never saw *this* coming. As the ovation continues, I'm still trying to understand what's happening. What is this feeling? It's not pride or satisfaction or even happiness. Could it be redemption? The saltwater in my eyes seems to think so. Eventually the clapping fades away, but this feeling profoundly remains.

THE LESSON OF THE MOVIE OF THE LIFE

In my opinion, Atari: Game Over is a great movie. Its fresh perspective reveals a number of things that have been hidden for a long time. I like it because it's an unfolding tale of human foibles and missteps that makes me feel good. And every time I see myself getting choked up at the dig, I choke up all over again.

It turns out the legend of the burial, like so many things in life, is true but not accurate. It was never about burying millions of E.T. game cartridges. The majority of the salvaged bounty were hits; Top sellers like Defender, Centipede and Yars' Revenge. There were consoles and peripherals, too. This was clearly a warehouse dump, not an E.T. graveyard. So maybe it didn't make sense to bury millions of E.T. games just to hide their corporate shame after all. But then again, what sense does it make to create a legend around it?

And speaking of shame…

I've always believed in humor as a tool, and it has served me well in many ways. I find hammers useful also, though I occasionally hit my thumb. Sometimes I use humor to hide from myself. Standing in that theater, surrounded by all the excitement and hoopla, a revelation came to me. I saw what I'd hidden from myself.

I realized during the ovation that I had carried a lot of pain for many years. Thinking I worked so hard to create such a disaster, it weighed on me. All the layers of humor and ownership had built a thick protective shell. In one remarkable moment, that shell shattered and fell away. To my surprise, what came rushing in was not pain. I felt a warm wave of compassion and appreciation. For the first time ever, I acknowledged all the shame I have carried, and my worst fears were not realized. Instead I experienced a sacred relief.

Tears are emotional language. They speak our truth when words cannot. I could never describe the impact of feeling so understood and absolved. But in that moment, my tears spoke eloquently.

CHAPTER 22
AFTERGLOW

It's late afternoon in the desert and the droning rhythm of heavy machinery has ceased, leaving a rich vein of historical relics running through endless piles of useless waste. The anthropologists are here to help determine which is which.

The storm is completely gone, giving way to a beautiful calm. I'm standing on the periphery of the Alamogordo city dump, chatting enjoyably with Zak while the film crew captures a few last shots. A vast horizon stretches out before us, enhanced considerably by a spectacular sunset. I'll say one thing for the New Mexico desert, it really knows how to accessorize.

It occurs to me how fitting it is that this incredibly turbulent, chaotic and ultimately productive day leads to such a satisfyingly tranquil conclusion. So it's been with my Atari experience. After my long post-Atari sojourn, I spend my days helping others settle the chaotic storms in their minds. I've arrived at a profound inner peace, finally reestablished in work that's authentically me.

As the evening sky colors its way into dusk, I'm reminded of an ending for a book I've always wanted to use:

> "And as the sun slowly sinks into the east,
> we notice the Earth is rotating in the wrong direction."

But I'm facing westward and feeling more sincere, so that won't do. I'd rather just share from my heart, which is wide open.

Atari was a kaleidoscope of dreams, a cornucopia of outrageous goings on. In addition to being an incredibly fun, rewarding and valuable experience, it was one of the most generous learning opportunities of my life. I'm still realizing lessons from my time at Atari. It was truly an amazing place to be. And when it was over, I didn't just leave Atari, I had to recover from it. Because if you were an Atari insider, tuned in and committed to what you were doing, it was a life changing experience. It certainly was for me. I'm still feeling the magic.

Life has a funny way of coming full circle. After three decades the video game industry is back to making simple games for smaller screens. I've come full circle too. Back then I entertained techies and nerds. Now I help make their lives better.

I've tried to honor a fifteen-year-old's commitment to creating a life worth living. Now I'm loving life as a psychotherapist because I feel like I was made for this. Same as I felt at Atari. How blessed am I to be able to say this twice in my life?

I set out to make a difference by producing things to delight, enrich and generate social discourse. I hope I succeeded.

After working at Atari for a while, I found myself talking with a realtor about finding a house. As a getting-to-know-you exercise, the realtor asked me: "What do you want written on your tombstone?"

My instant response was, of course, "He's not here yet."

But after reflecting a bit, I came up with something more genuine to my life's ambition, "Howard was an artist. His medium was life."

From the day I was born, my life has been a quest to figure a few things out: *Who am I? Where am I heading? What do I really want? What shall I do?* Though I accomplished a great deal in my first 23 years, these questions haunted me… until I arrived at 1272 Borregas Ave. Atari was a rich bounty in my life, but the most precious gift I received was answers.

Atari was not my birthplace, it was simply where my life began.

AFTERWORD

I've known Howard through many of his iterations. Like a cat with nine lives, I've lived with him through a bunch of his. And no matter what he was trying to do, documentary filmmaker, screenwriter, computer engineer (again), and finally, therapist, for the 20 years we've been friends I continually asked him, "Where's the Atari Book?"

Well, here it is. And it was worth the wait. I love this book. As a non-technical person, or as Howard says "non-Nerd", it's exactly what I hoped it would be, an accessible, funny, whimsical, accurate telling of the birth of the gaming business from a guy who was there and actually remembers it. Howard has the sharpest mind and sharpest wit of anyone I know, and even though I work very hard never to laugh at anything he says because I know it secretly drives him crazy, it's on full display here.

Way to go Howard. You did it. The ultimate computer gaming book. You want to know where those incredible games you play now originally came from. They came through Howard.

Bob Saenz
Screenwriter & Author.

AUTHOR

Howard Scott Warshaw is the most famous person you've never heard of. He's an artist, technologist, creator and healer, but first and foremost he is a communicator. Holding master's degrees in Counseling Psychology and Computer Engineering, his career accomplishments include Video Game pioneer, MoMA artist, celebrated software developer, award winning film producer, author, educator and columnist. These days Howard enlists his eclectic skill set in the service of others as a psychotherapist in California's Silicon Valley, where he specializes in the issues of hi-tech leaders and the super-intelligent. He loves cultivating new skills and finding creative ways to apply them. Howard is a complex person who can be summed up in five words: Passion with a Balanced Perspective.

Also by Howard:

Inspired Therapist
Conquering College
The Complete Book of Pan
Once Upon Atari DVD Documentary Series

www.ingramcontent.com/pod-product-compliance
Lightning Source LLC
Chambersburg PA
CBHW062242300426
44110CB00034B/1169